THE SILENT UNFOLDING

Metamorphosis as the Universe's Hidden Language

by

Calyx Drevorn

First edition

Printed on acid-free paper

For every living thing that has ever endured
a season of not knowing what it was becoming.

> *The butterfly does not remember being a caterpillar. Yet it could not exist without having been one. This is not loss. It is the deepest kind of continuity.*
>
> — *Calyx Drevorn*

> *All things flow; nothing stands still.*
>
> — *Heraclitus, Fragment 91 (c. 500 BCE), as recorded by Simplicius*

Contents

Foreword
Preface: Why This Book Now
How to Read This Book
A Note on Sources and Methodology
Introduction: The Grammar of Change

Part I The Biology Beneath the Biology
Chapter 1 The Creature That Dissolves Itself
Chapter 2 Beyond the Butterfly
Chapter 3 The Metamorphic Genome

Part II The Cosmic Blueprint
Chapter 4 Stars Are Creatures Too
Chapter 5 Quantum Metamorphosis
Chapter 6 The Dark Chrysalis

Part III The Human Dimension
Chapter 7 The Brain That Remakes Itself
Chapter 8 Grief, Loss, and the Cocoon You Did Not Choose
Chapter 9 Who Are You After?
Chapter 10 Civilizations in the Chrysalis

Part IV The Philosophical Core
Chapter 11 Is Change Real?
Chapter 12 The Ethics of Becoming
Chapter 13 Does Metamorphosis Have a Direction?

Part V The Future of Metamorphosis

Chapter 14 Engineering Emergence
Chapter 15 Artificial Intelligence as Metamorphic Intelligence
Chapter 16 Can Humanity Collectively Metamorphose?

Conclusion: The Art of Metamorphic Living
Afterword
Acknowledgments
Glossary of Key Terms
Notes and References
Recommended Reading
About the Author

Foreword

There is a moment in the life of every inquiry when a question that seemed limited suddenly reveals itself to be universal. For the author of this book, that moment came not in a laboratory or a library, but on an evening walk through a field in early autumn, when the air carried the faint smell of decomposing leaves and a single moth, barely visible in the dimming light, was completing its emergence from a cocoon attached to a dried stem of goldenrod.

What Calyx Drevorn witnessed that evening was, by biological measure, unremarkable. Moths emerge from cocoons every second of every day across every continent where they exist. The transformation from larval caterpillar to winged adult is so common in nature that it barely merits a footnote in most surveys of insect biology. And yet the author stood watching for nearly twenty minutes, unable to walk away, gripped by something that exceeded the merely aesthetic.

The grip, this book argues, was not accidental. It was recognition.

We respond to metamorphosis with disproportionate feeling because, on some level we rarely articulate, we know ourselves to be metamorphic creatures. We have all lived through the experience of one version of ourselves ending and another beginning, often without having chosen the transition and often without knowing, while inside it, whether anything coherent would emerge on the other side. The cocoon resonates because we have, in our own ways, inhabited it.

This book is the most ambitious attempt yet made to examine metamorphosis not as a biological specialty but as a governing principle of reality. Drawing on developmental biology, astrophysics, quantum mechanics, depth psychology, anthropology, and process philosophy, Calyx Drevorn builds a case that transformation of the radical, reorganizing, identity-altering kind is not an exception in nature. It is nature's primary method.

The argument unfolds in five movements. The first grounds us in the strangeness of biological metamorphosis, revealing details about what actually happens inside a chrysalis that most educated readers will find genuinely startling. The second scales the metamorphic pattern outward to the cosmos, tracing the same structural logic in stellar evolution, quantum mechanics, and the unseen architecture of dark matter. The third turns inward, applying the framework to the human mind, to grief, to personal identity, and to the transformation of cultures and civilizations. The fourth confronts the deepest philosophical questions the framework raises: Is change real? What do we owe our former selves? Does the pattern imply direction or meaning? The fifth looks forward, examining the metamorphic implications of genetic engineering, artificial intelligence, and the question of whether humanity as a whole can navigate the transformation now underway.

What makes this project unusual is not its scope, ambitious as that scope is, but its governing insistence on intellectual honesty. Drevorn does not flatten the evidence to fit a thesis. Where the science is settled, it is presented as settled. Where it is contested, the contest is acknowledged. Where the argument crosses from empirical ground into philosophical speculation, the crossing is marked explicitly. Readers will not find here the kind of overreaching that afflicts so much cross-disciplinary writing, where the excitement of a unifying idea overwhelms the rigor that should restrain it.

They will find, instead, a book that takes both the science and the wonder seriously, and that earns its larger claims by attending carefully to the smaller ones.

The title, The Silent Unfolding, comes from a phrase that recurs in the biological literature on holometabolous development, referring to the period inside the chrysalis when all outward signs of activity have ceased and yet the most radical reorganization is underway. It is a phrase that resonates far beyond entomology. The most transformative processes are often invisible from outside. The most profound changes happen in silence.

That observation, which sounds at first like a truism, turns out on examination to have consequences that reach into physics, into psychology, into the history of civilizations, and into the question of what it means to live a human life with intelligence and courage. By the time you have finished this book, you will understand why.

Preface

Why This Book Now

The moth I watched that autumn evening emerged slowly, as they always do. The wings, which had been folded and wet and useless, spread incrementally over the course of perhaps fifteen minutes, stiffening as hemolymph pumped through the venation, the pattern of the forewings resolving from a crumpled abstraction into something precise and intentional-looking. When the process was complete, the creature pumped its wings twice, as though testing them, and then was still.

I did not know then what I know now about what had preceded that emergence. I did not know that the caterpillar from which this moth had developed had, inside the cocoon, dissolved the majority of its own tissues into a biological fluid, digesting the very structures that had constituted its former self. I did not know that the cells that built the wings had been present since the egg stage, dormant and patient, waiting for exactly this dissolution to begin their work. I did not know that the caterpillar's own immune system had attacked these future-building cells as foreign invaders before the transformation's chemistry finally overcame the resistance.

I did not know any of this. But I think some part of me understood it.

I had, in the two years preceding that autumn, been through a period that I can only describe, now that I have the vocabulary, as my own chrysalis event. A long-term relationship had ended. A career path I had pursued for nearly a decade had revealed itself as wrong for me in a way that could not be corrected by adjustment or persistence. Several of the beliefs through which I had organized my understanding of myself and the world had not survived close examination. I was, in the language that will become familiar over the course of this book, mid-dissolve.

I do not tell you this because this is a book about my life. It is not. I tell you this because the questions that animate the pages that follow did not originate in an academic interest in biological transformation or a desire to write a synthesis of natural philosophy. They originated in the urgent, practical need to understand what was happening to me, and to know whether what looked from the inside like destruction might, from some wider vantage, be something else.

What I found, across seven years of research, reading, and conversation with scientists and philosophers and ordinary people who had navigated profound personal transformation, was more than I had expected. I found that the structure of metamorphosis, understood not merely biologically but as a pattern recurring across scales of reality from the subatomic to the cosmological, is one of the most rigorously described and least understood phenomena in the history of human inquiry. I found that specialists in each of the relevant fields, biologists and physicists and psychologists and historians and philosophers, had been examining the same underlying pattern from different angles without, in most cases, knowing that the others were doing so. And I found that the pattern, when assembled into a single view, had implications that reached well beyond any single discipline and into the question of how a human being ought to live.

This book is the product of that seven-year investigation. It is written for general readers rather than specialists, though I hope specialists in each of the relevant fields will find the treatment of their own discipline sufficiently rigorous to be worth their time. It is written in the conviction that the most important questions are those that cross disciplinary boundaries, and that the courage to follow an idea wherever it leads, even when it leads outside the domain where you feel competent and secure, is itself a kind of metamorphic practice.

A word about the structure. The book moves, as I have said, like metamorphosis itself: from solidity to dissolution to reorganization to emergence. The early chapters are deliberately grounding, presenting the biological facts in enough detail that when the argument scales outward to cosmology and inward to psychology, you will have a concrete anchor. The middle chapters are the most demanding, because they require holding complexity without premature resolution. The final chapters are, I hope, genuinely illuminating rather than merely optimistic.

You may, at certain points in the middle of this book, feel something analogous to the disorientation that characterizes the chrysalis phase. This is not a flaw in the design. It is the design. Real understanding, like real transformation, requires a period of not-yet-knowing. I ask you to trust that period, as the caterpillar appears to trust it, and to stay in the book until you reach the other side.

The moth I watched that evening flew before the light was entirely gone. I watched it until it disappeared, which was almost immediately, into the dark. I have been thinking about it, in one way or another, ever since.

How to Read This Book

This book can be read straight through from beginning to end, and that is the reading sequence for which it was designed. The argument builds across the five parts, and the later chapters depend on the earlier ones for their fullest effect. That said, different readers come with different urgencies, and a few notes on structure may help you find your way into the material at the point where it speaks to you most directly.

The five parts correspond to five scales of metamorphic reality. Part One examines the phenomenon at the biological level, where the evidence is most concrete and the science most settled. Part Two scales outward to cosmology and physics, making the case that the same structural logic visible in the chrysalis is visible in the life and death of stars. Part Three turns inward to the human scale, treating the brain, the psyche, grief, identity, and culture as metamorphic systems. Part Four is philosophical, confronting the hardest conceptual questions the framework raises. Part Five is prospective, examining the metamorphic transformations that are now underway in biology, technology, and civilization.

For the Reader Who Comes from Science

Begin with Part One and read the biological chapters carefully. Then move to Part Two, where the scientific claims are more speculative. The chapters in Part Two are written with full acknowledgment of where the argument crosses from established science into philosophical interpretation, and I would ask scientific readers to hold that distinction as they read. The goal of Parts Two and Five is not to misuse science but to think with it, to allow empirical findings to illuminate questions that lie beyond the empirical domain.

For the Reader Who Comes from Psychology or Humanistic Study

Begin with Part Three, then return to Parts One and Two. The psychological and cultural chapters are designed to stand on their own, but they will be considerably enriched by the biological and cosmological framework. Part Four may be particularly compelling for readers with a background in philosophy or literary theory.

For the Reader in the Middle of a Transition

Chapter Eight was written with you in mind. But so, in a sense, was the entire book. If you are currently in what this book will call the chrysalis phase of a personal, professional, or relational metamorphosis, I would suggest beginning with the Introduction, moving to Chapter Eight, and then reading Chapter Sixteen before returning to the beginning. The full argument will make more sense to you after you have seen where it is going.

Whichever path you take, the book's central claim will meet you: that metamorphosis is not what happens to some things sometimes. It is what reality does, at every scale, consistently and without exception. Understanding this changes not only how you think about nature but how you live inside your own transformations.

A Note on Sources and Methodology

This book draws on primary scientific literature, peer-reviewed research, and the published work of specialists across a wide range of disciplines. Where a claim is specific and verifiable, a citation is provided in the Notes section at the back of the book. Where I am summarizing a broad scientific consensus, I have indicated that in the text. Where I am making a philosophical argument that goes beyond what the science directly supports, I have said so explicitly.

I am not a biologist, a physicist, a neuroscientist, or a professional philosopher. I am a researcher and writer with a background in the history and philosophy of science, and the writing of this book required me to learn, at a working level, material in all of those fields. I have been helped enormously by conversations with researchers in each area, and I have tried to represent their work accurately. Where I have made errors, they are mine and not theirs, and I welcome corrections.

Three methodological commitments shape the book throughout. First, I have tried to let the evidence complicate the thesis rather than simplify it. The argument that metamorphosis is a universal organizing principle is a large claim, and large claims require rigorous scrutiny. I have tried to engage that scrutiny honestly rather than avoiding it. Second, I have tried to maintain a clear distinction between what is known, what is hypothesized, and what is speculative. Third, I have included no data that I cannot trace to a primary source, and no anecdotes that I cannot verify. The book contains no composite characters, no invented scenarios, and no fabricated statistics.

The philosophical sections of the book are more openly interpretive, as philosophical writing necessarily is. But I have tried, even there, to distinguish between arguments that follow from the evidence and positions that represent my own judgment, held with appropriate tentativeness, about what the evidence implies.

Introduction

The Grammar of Change

Consider three events that appear to have nothing in common.

In the Milky Way galaxy, roughly six thousand years ago, a massive star ran out of hydrogen fuel. The outward pressure of nuclear fusion, which had held the star's structure in equilibrium for ten million years, ceased abruptly. The star's iron core collapsed in approximately a quarter of a second, compressing a mass greater than our sun into a sphere roughly twenty kilometers across. The rebound shock wave expelled the outer layers of the star into space at ten percent of the speed of light. In the process, temperatures reached one hundred billion degrees Celsius, high enough to forge elements that could not have been created in any previous stellar phase. The explosion was visible from Earth with the naked eye for months, burning as bright as a full moon in the daylight sky. The remnant today is the Crab Nebula: a six-light-year-wide cloud of expanding gas and a spinning neutron star at its center, rotating thirty times per second and emitting beams of radiation that sweep past Earth like a lighthouse beam.

In a field in Peru, in late spring, a hawk moth caterpillar of the species Manduca sexta attached itself to a plant stem and began spinning a silken structure around its body. Over the next two weeks, something happened inside that structure that does not fit comfortably within the usual language of biological development. The caterpillar did not grow wings. Instead, it secreted digestive enzymes that dissolved the majority of its own tissues, including its muscles, its fat body, and most of its gut, into an undifferentiated biological fluid. From this fluid, guided by cells that had been dormant since the egg stage, an entirely different body plan was constructed: a body with compound eyes, a six-legged adult form, wings, and a proboscis capable of reaching deep into flowers. The process required the coordinated death of billions of cells and the birth of billions of others. When it was complete, the structure that had housed all this reorganization split open and a moth emerged that shared almost none of its predecessor's physical architecture.

And in a hospital room in London, in 2007, a forty-three-year-old woman named Diane lay in a bed recovering from a bilateral mastectomy following a diagnosis of stage three breast cancer. She had been a professional cellist before the diagnosis. She would not play again for two years, not because of physical limitation but because, as she later described it, she could not be certain, for a long time, that the person who had played the cello was still present. The person who woke each morning in the months following her surgery felt continuous with her former self in some respects but not others. Her values had shifted. Her priorities had reorganized. Relationships that had seemed permanent felt provisional; relationships she had treated as secondary had moved to the center of her life. She described the experience, in an interview conducted for this book, as feeling not like recovery but like becoming someone she had not met before.

Three events. A supernova, a moth, a woman. What connects them?

The answer this book proposes is not metaphorical or impressionistic. It is structural. All three events follow the same logic: a stable organized system encounters conditions that exceed its capacity for incremental adaptation; the system undergoes a period of radical dissolution in which its previous structure is largely dismantled; during this dissolution, elements that had been latent or dormant become active; from these elements, a new organization emerges that is continuous with the old in some respects and radically discontinuous in others; the new organization possesses capacities that the old one did not.

This is the grammar of metamorphosis. And this book's central claim is that this grammar, this sequence of dissolution, reorganization, and emergence, is not the special property of a particular class of organisms or a particular category of human experience. It is a pattern written into the structure of reality at every scale we have been able to examine, from the behavior of quantum particles to the transformation of civilizations.

What This Claim Is and What It Is Not

Before proceeding, a clarification is essential. The claim that metamorphosis follows a universal structural grammar is not a claim that everything transforms in the same way, at the same rate, or with the same outcome. The star, the moth, and the woman underwent profoundly different processes in profoundly different timeframes for profoundly different reasons. What they share is not the mechanism but the logic: the sequence of phases through which radical transformation moves.

Nor is this a claim that all transformations succeed. The vast majority of metamorphic events in nature fail. Most caterpillars do not survive to emergence; most are eaten, infected, or killed by cold. Most stars that undergo supernova do not produce stable remnants; many collapse directly into black holes from which nothing escapes. Most human beings who enter profound personal transformation do not complete it in the creative sense; many remain in the dissolution phase indefinitely, or return to a form not significantly different from the one they left.

What the universal grammar of metamorphosis offers is not a guarantee but a map. It describes the structure of the territory, not a route through it. Understanding the structure does not make the journey easier, but it may make it less terrifying, because the dissolution that feels like ending may, when mapped against the larger pattern, be recognizable as middle.

Three Concepts That Will Carry the Argument

Three concepts will appear repeatedly throughout this book, and it is worth introducing them now.

The first is the imaginal cell. In holometabolous insects, those that undergo complete metamorphosis, the cells that build the adult body are called imaginal cells or imaginal discs. They are present from the earliest larval stage but remain dormant throughout it, expressing neither their identity nor their function. During the dissolution phase inside the chrysalis, when the caterpillar's body is largely liquefied, these cells activate and begin organizing the new body. The caterpillar's own immune system initially treats them as foreign invaders and attacks them; they survive only because they carry chemical markers that protect them from complete destruction, and because at a certain point in the transformation's chemistry, they begin connecting with one another and generating a signal environment that the immune system cannot suppress.

The imaginal cell is, throughout this book, a metaphor as well as a biological fact. It stands for the elements within any system that carry the blueprint for a future organization that the current organization cannot yet accommodate: dormant potentials in a human psyche, innovative minorities within a culture, new physical laws that operate within conditions where old ones have broken down. Recognizing imaginal cells, in any system, is one of the skills that metamorphic literacy makes possible.

The second concept is the chrysalis paradox. The chrysalis, the structure within which metamorphosis occurs, is from outside almost indistinguishable from a dead thing. It does not move. It does not eat. It does not respond to stimulation in the ways that living organisms characteristically respond. Everything about it says: this has ended. The paradox is that the most intense biological activity in the insect's entire life cycle is occurring inside it. The chrysalis paradox describes any situation in which maximal internal activity coincides with minimal external expression, and in which the absence of visible progress is itself evidence of transformation.

The third concept is metamorphic literacy: the capacity to recognize a metamorphic event when you are inside one, to distinguish dissolution from mere deterioration, and to respond appropriately to each phase of the transformation. This book is, among other things, an attempt to develop that literacy. It will not make the metamorphic phases of your own life less painful or disorienting. But it may make them more legible.

A Map of What Follows

Part One begins where the evidence is most concrete: in the biology of organisms that undergo complete metamorphosis. The three chapters of Part One cover what actually happens inside a chrysalis, how metamorphosis is distributed across the kingdoms of life, and what the molecular biology of transformation reveals about its deep evolutionary origins. These chapters are the foundation; the later argument cannot stand without them.

Part Two makes the largest single claim of the book: that the structural grammar of metamorphosis is visible not only in biology but in the physical universe. The chapters on stellar evolution, quantum mechanics, and dark matter are the most speculative in the book, in the sense that they require crossing from empirical description to philosophical interpretation. I have tried to make those crossings clearly marked.

Part Three is where the argument becomes personal. The chapters on neuroplasticity, grief, identity, and civilizational transformation draw on the biological and cosmological framework to illuminate dimensions of human experience that have been poorly served by the languages currently available for describing them.

Part Four is philosophical in the strict sense: it examines the logical and ethical implications of the metamorphic framework. These chapters are the most demanding and, I believe, the most rewarding.

Part Five is prospective. It examines what human beings are now doing with the power to redesign biological metamorphosis, what artificial intelligence represents in metamorphic terms, and what would have to be true for humanity to navigate the transformation now underway.

The book ends where it begins: with a moth, and with the question of what it means to have watched one.

I

The Creature That Dissolves Itself

What actually happens inside a cocoon

Most people, if asked to describe what happens inside a chrysalis, would say something like: the caterpillar grows wings. This is understandable. It is also, as a description of the actual process, almost entirely wrong.

Wings do not grow from a caterpillar. A caterpillar does not grow into anything, in the sense that we normally mean when we use that phrase. What happens inside a chrysalis is something far stranger, and the strangeness matters because it is not merely biological trivia. It is the first and most concrete demonstration of the central claim of this book: that the most radical transformations do not proceed through gradual accumulation but through a phase of fundamental dissolution from which something new emerges.

Here is what actually happens.

The Biology of Dissolution

When a caterpillar of a holometabolous species, those that undergo complete metamorphosis, has reached the appropriate developmental stage and received the appropriate hormonal signals, it stops feeding and begins to change its behavior. In Lepidoptera, the moths and butterflies, the caterpillar typically attaches itself to a substrate and molts for the final time, revealing a hardened outer layer that becomes the chrysalis or cocoon. Inside this structure, the caterpillar secretes enzymes that trigger a process called histolysis: the systematic breakdown of larval tissues.

Histolysis is not selective in any meaningful sense. The caterpillar's muscles, which have carried it through weeks of feeding and movement, dissolve. The fat body, which stores the metabolic reserves that will fuel the entire transformation, dissolves. The digestive tract, which has processed the enormous quantities of plant material required to grow from a few milligrams to several grams in a matter of weeks, dissolves. The result is not a caterpillar that has died but something more difficult to categorize: a living biological system that has dismantled nearly all of the structures through which it previously expressed its identity as a caterpillar, and that is now, in metabolic terms, something between a liquid and a solid.

The fluid inside a chrysalis at this stage of development is alive. It maintains biochemical activity, temperature regulation, and hormonal signaling. But it has no muscles, no functional digestive system, no coherent body plan. The biologist Lynn Riddiford of the University of Washington, who has spent four decades studying the hormonal regulation of insect development, has described the interior of a developing chrysalis as resembling a rich biological soup: not dead, not alive in the usual sense, but suspended between organizational states.

From this soup, a new body is built.

Imaginal Discs: The Blueprint That Survived

The structures that build the adult body are called imaginal discs, from the Latin imago, meaning the adult form of an insect. They are clusters of undifferentiated cells, numbering on the order of ten thousand to fifty thousand cells per disc in a typical Lepidopteran species, that are present in the larva from the earliest stages of development. In the caterpillar, they sit dormant, expressing none of their adult potential, tucked within the larval tissue like sealed envelopes containing instructions that have not yet been opened.

There are typically ten to twelve pairs of imaginal discs in a holometabolous insect larva. Each disc is genetically programmed to produce a specific adult structure: the wing disc will produce a wing, the leg disc a leg, the eye disc the compound eye. During the larval stage, these discs grow slowly, dividing to increase their cell count while remaining in an undifferentiated state. They are, in the language of developmental biology, committed but not yet differentiated: they have received the signal that specifies their eventual identity but have not yet begun to express it.

When histolysis begins and the larval body dissolves around them, the imaginal discs do not dissolve with it. They are protected by their cellular characteristics from the enzymes that break down larval tissue. And it is from these discs, drawing on the metabolic resources released by the dissolution of the larval body, that the adult insect is constructed.

The wing disc of a Monarch butterfly, Danaus plexippus, for example, begins as a small cluster of undifferentiated cells in the caterpillar larva. By the time it has finished its work inside the chrysalis, it has produced a structure with approximately two hundred thousand cells organized into a precise architectural pattern of veins and membrane that, when correctly expanded with hemolymph and dried, will support flight at speeds of up to thirty-five kilometers per hour and navigate, across thousands of kilometers, to overwintering sites in the mountains of Michoacan, Mexico, that individual butterflies have never visited.

The precision of what the imaginal disc produces, from such modest origins, remains one of the most remarkable phenomena in developmental biology.

The Immune System Attacks Itself

There is a dimension of metamorphosis that receives almost no attention in popular accounts and that seems, when you first encounter it, almost too metaphorically convenient to be true. It is true.

The caterpillar's immune system attacks the imaginal cells.

Because imaginal cells differ biochemically from the larval cells that surround them, the caterpillar's immune surveillance system registers them as foreign and mounts an immune response against them. In the early stages of the larval period, the imaginal cells are small enough in number and their chemical signatures subtle enough that this immune attack does not eliminate them. But as they grow and as the transformation chemistry intensifies, the conflict between the immune system and the imaginal cells becomes a genuine competition, one that the imaginal cells win, ultimately, not by defeating the immune system directly but by connecting with each other.

Research published in 2002 by biologists at Georgetown University demonstrated that as imaginal cells accumulate in developing pupae, they begin exchanging chemical signals that function as a kind of recognition system. Cells that share the imaginal disc identity begin to cluster, forming increasingly large masses that the immune system becomes less capable of treating as individual foreign targets. When enough imaginal cells are connected and coordinating with each other, they generate a biochemical environment that effectively overrides the immune suppression.

This sequence, in which elements carrying the blueprint for a new organization are initially suppressed by the existing system before eventually reaching a critical mass at which suppression becomes impossible, is one that will appear again in this book in contexts far removed from entomology.

The Question of Memory

One of the most philosophically provocative questions raised by metamorphosis is whether anything that was learned or experienced by the caterpillar persists into the adult. This is not a trivial question. The caterpillar and the adult butterfly are, from a structural perspective, almost entirely different organisms. The nervous system is substantially reorganized during metamorphosis, with most of the larval nervous system dissolving and a new adult nervous system developing from neuroblasts. The sensory systems are entirely replaced: the caterpillar's simple eyes, which can distinguish light from dark but not form, are replaced by the compound eyes of the adult, capable of detecting polarized light and ultraviolet wavelengths invisible to humans.

Given this reorganization, it would be reasonable to assume that anything the caterpillar learned would be entirely erased.

In 2008, a team of researchers at Georgetown University led by Douglas Blackiston and Martha Weiss published a study in the journal PLOS ONE that tested this assumption with one of the most elegant experimental designs in the recent history of behavioral biology. They trained tobacco hornworm caterpillars, Manduca sexta, to avoid a particular odor by associating it with a mild electric shock. The caterpillars successfully learned the avoidance behavior. The researchers then allowed the caterpillars to pupate and emerge as moths, and tested the moths for memory of the trained avoidance response.

More than 77 percent of the adult moths that had been trained as early-instar larvae retained the avoidance behavior. Memory, or something that functions as memory, had survived not merely the dissolution of the larval body but the radical reorganization of the nervous system.

The implications of this finding are still being worked out. It appears that some neural circuits in the larval nervous system survive metamorphosis relatively intact and are incorporated into the adult system, and that these circuits can encode learned information across the transformation. The mechanisms are not yet fully understood. But the finding establishes something that matters beyond neuroscience: that metamorphosis is not simply destruction followed by reconstruction from scratch. It is reorganization that preserves, selectively, elements of the original while building structures that the original could not accommodate.

Who Is the Butterfly?

The philosophical question that the biology of metamorphosis raises most insistently is this: in what sense is the butterfly the same organism as the caterpillar from which it emerged?

The genetic identity is continuous: the butterfly and the caterpillar share the same genome. The chronological continuity is unbroken: the butterfly is older than the caterpillar was at the moment of pupation, and the developmental process from one to the other was continuous in time if not in structure. Some neural information, as the Blackiston study suggests, may be preserved. In these respects, there is a case for identity.

But the architecture is almost entirely different. The body plan is different. The sensory systems are different. The dietary requirements are different: the caterpillar consumed plant material; the adult butterfly, if it feeds at all, consumes nectar. The behavioral repertoire is different. The habitat is different. The ecological role is different.

A philosopher of personal identity presented with this case and asked whether the butterfly is the same organism as the caterpillar would face a genuine puzzle. The case for continuity is real but partial. The case for discontinuity is equally real. The most honest answer is probably that the binary question, same or not the same, is the wrong question. Metamorphosis produces something that is neither straightforwardly the same as what preceded it nor straightforwardly different: it produces a continuation that is also a transformation.

We will return to this puzzle in Chapter Nine, when we examine what it means for human beings who undergo radical personal transformation. For now, it is enough to notice that biology itself, in the most extreme case of transformation it has produced, does not resolve the question cleanly. The caterpillar and the butterfly are connected by a process that is neither simple continuity nor simple replacement. They are related by metamorphosis. And metamorphosis, as this book will argue at length, is a category that requires its own logic.

Why This Matters Beyond Biology

The account of metamorphosis given in this chapter is, in its essentials, biology. But the reason it belongs at the beginning of a book that will go on to discuss cosmology, psychology, and the future of civilization is that it establishes the foundational pattern in its most concrete and scientifically grounded form.

The pattern is this: the capacity for radical transformation is not something that is added to a system from outside. It is present within the system from the beginning, encoded in structures that remain dormant while the original organization persists. The transition from one form to another requires a phase of dissolution in which the original structure is not incrementally modified but substantially dismantled. The dissolution is not random but directed: it releases the resources and the conditions necessary for the new organization to emerge. The existing system initially resists the new organization, registering it as foreign. But at a certain point, the new organization reaches a critical mass at which the resistance can no longer suppress it.

This pattern will appear again, and then again, in contexts that have nothing biological about them. It will appear in the physics of phase transitions, in the evolutionary history of the universe, in the neuroscience of personal transformation, in the historical dynamics of civilizational change. Each new appearance of the pattern will be different in its details. But the logic will be the same.

We are watching, in the biology of metamorphosis, the universe's most carefully executed demonstration of how radical change works. The caterpillar is dissolving. The imaginal cells are activating. Something entirely new is being built from the resources of something that has let itself come apart.

It has been doing this, in one species or another, for roughly three hundred million years.

II

Beyond the Butterfly

Metamorphosis across the web of life

In September 2000, the journal Nature published a four-page paper by Toshiyuki Nakagaki and colleagues from Hokkaido University in Japan. The paper described an experiment in which a plasmodial slime mold, specifically Physarum polycephalum, had been placed at one entrance of a maze and a food source placed at another. The slime mold, which has no nervous system, no neurons, and no brain of any kind, found the shortest path through the maze within four hours.

The paper generated considerable attention, but it was a follow-up study, published in Science in 2010, that became iconic. Nakagaki's team placed food sources on a surface at positions corresponding to the major cities of the Tokyo metropolitan area. They placed the slime mold at the position corresponding to Tokyo. Without any instruction or programming, the organism extended nutrient-transporting tubes that, when the growth pattern had stabilized, closely resembled the actual configuration of the Tokyo rail network, a system that had taken human engineers decades to design and refine for efficiency.

The slime mold had, without a nervous system, without planning, without memory in any conventional sense, reproduced an optimized transport network through a process of self-organization.

This experiment has nothing directly to do with metamorphosis in the biological sense used in Chapter One. But it is the right place to begin this chapter because it illustrates the central point that the chapter will develop: the biological strategies we associate with the transformation of caterpillars into butterflies are not the specialized invention of one group of insects. They are expressions of organizing principles that appear, in various forms and through various mechanisms, across every major branch of life.

The Spectrum of Biological Transformation

Biologists distinguish between two broad types of metamorphosis in insects. Hemimetabolism, or incomplete metamorphosis, describes the developmental pattern of species such as grasshoppers and dragonflies, in which the young hatch as nymphs that resemble miniature adults and gradually acquire adult characteristics through a series of molts. There is real transformation here, but it is incremental rather than radical: the nymph is recognizably the same kind of creature as the adult it will become.

Holometabolism, or complete metamorphosis, is the pattern described in Chapter One: egg, larva, pupa, adult. Between the larval stage and the adult stage lies the pupal phase, in which the radical reorganization of the body occurs. Holometabolism evolved at least once and possibly multiple times in the history of insects, and it has been enormously successful: approximately 85 percent of all insect species, representing somewhere in the range of one million described species and an estimated four to eight million species in total, are holometabolous.

The scale of this success is worth pausing on. If holometabolism were a minor biological quirk, an adaptive strategy useful in a narrow range of ecological contexts, it would not have proliferated so extensively. Its dominance among insects suggests that the separation of body plans between larval and adult stages, with a radical reorganization between them, offers adaptive advantages that incremental development cannot match. Larvae and adults can exploit entirely different ecological niches, food sources, and habitats, effectively doubling the ecological space available to the species. The cost, a period of radical dissolution and reorganization during which the organism is vulnerable and cannot feed or move effectively, is apparently worth paying.

Transformation in the Sea

Marine biology offers some of the most dramatic examples of metamorphosis outside the insect world. The metamorphosis of certain species of sea cucumber, members of the class Holothuroidea, involves a strategy that has no analogue in terrestrial biology: voluntary evisceration.

When a sea cucumber is sufficiently stressed, by temperature shock, predation, or other environmental insults, it ejects its own internal organs through the body wall. The respiratory trees, the digestive system, and portions of the reproductive system are expelled into the surrounding water. What remains is, anatomically, a remarkably simplified creature: essentially a body wall surrounding an empty cavity. Over the following two to five weeks, the sea cucumber regenerates its lost organs entirely, producing a new digestive system and new respiratory structures from remaining tissue.

This is not, technically, metamorphosis in the sense of a transition from one life stage to another. But it is radical enough that the regenerated organs are not simply copies of the originals. The regenerated tissue is reorganized and, in some cases, produces structures with slightly different characteristics than those that were expelled. The sea cucumber that emerges from the evisceration process is, in a meaningful physiological sense, not exactly the same organism that entered it.

More classically metamorphic are the echinoderms collectively, including sea stars, sea urchins, and sand dollars. The larvae of most echinoderm species look nothing like the adults they will become. Asteroid larvae, the precursors of sea stars, are bilaterally symmetrical, planktonic, and ciliated. The adult sea star is radially symmetrical, benthic, and covered in dermal ossicles. The transition between larval and adult form requires a reorganization of body symmetry, one of the most fundamental architectural properties of animal organization, that has few parallels in the animal kingdom.

The Axolotl and the Limits of Transformation

The axolotl, Ambystoma mexicanum, is a salamander native to the lake complex near Mexico City, and it offers one of the most unusual cases of metamorphic variation in vertebrate biology. Axolotls are neotenic: they retain larval features throughout their adult lives, including external gills and an aquatic lifestyle, even when sexually mature. They undergo what biologists call facultative metamorphosis: under certain hormonal conditions, they can transform into a terrestrial adult form, but in their natural environment they typically do not.

The axolotl is also celebrated for its regenerative capacity. An axolotl that loses a limb will regenerate it, complete with bones, muscles, nerves, and skin, within roughly two months. More remarkably, axolotls can regenerate portions of the heart, retinal tissue, and, to a limited degree, sections of the spinal cord and brain. The mechanism of this regeneration draws on dedifferentiation, the process by which fully differentiated adult cells revert to a less differentiated state before proliferating and redifferentiating to produce the regenerated tissue.

Dedifferentiation is, in effect, a partial metamorphosis: the cell abandons its adult identity, reverts toward something more like an embryonic state, and then re-specifies toward whatever identity is needed. The axolotl's extraordinary regenerative capacity depends on its ability to perform this local, targeted metamorphosis repeatedly throughout its life.

The regenerative biology of the axolotl is of intense interest to medical researchers for obvious reasons. But for the purposes of this chapter, it illustrates something conceptually important: metamorphosis is not a one-time event that happens once in a lifetime during a discrete developmental phase. It is a capacity that some organisms maintain throughout their lives, deploying it in response to damage, stress, or environmental change.

Bacterial Metamorphosis and the Ancient Origins of Transformation

The oldest forms of metamorphic transformation are microbial. Bacillus subtilis, a common soil bacterium, can, under conditions of nutrient deprivation or environmental stress, undergo sporulation: a process in which the bacterium partitions its genetic material into a new, smaller cell called a forespore, surrounds it with multiple protective layers, and then lyses, or breaks open, releasing the spore and effectively dying in the process of creating it.

The endospore that Bacillus subtilis produces is one of the most resilient biological structures known. It can survive temperatures above 150 degrees Celsius, ultraviolet radiation at lethal doses, and desiccation for extended periods. Spores recovered from the gut of bees preserved in Burmese amber, dating to approximately twenty-five million to forty million years ago, have reportedly been revived under laboratory conditions, though these claims remain contested in the scientific literature. Less controversially, spores have been revived after exposure to conditions that would kill virtually any other living cell.

The transition from vegetative bacterial cell to endospore is controlled by a highly conserved regulatory network, and the genetic components of this network show deep evolutionary relatedness to systems involved in developmental transitions in more complex organisms. This suggests that some of the molecular machinery of metamorphosis has ancient prokaryotic origins, and that the capacity for radical self-reorganization is not a late evolutionary invention but among the oldest biological strategies.

The Planarian Problem

Planarians are flatworms of the class Turbellaria, and they have a relationship with metamorphosis and regeneration that pushes the concepts to their conceptual limits. Planarians can be cut into pieces, and each piece, under appropriate conditions, will regenerate into a complete organism. A planarian cut transversely will produce a head fragment that grows a new tail and a tail fragment that grows a new head, including a new brain.

This is extraordinary enough. But research conducted at the Stowers Institute for Medical Research in Kansas City, published in 2013 by Michael Levin and colleagues, added a dimension that deepened the puzzle considerably. The researchers investigated whether the new brain produced in the regenerated planarian would adopt the original planarian's learned behavior. Working with a strain of planarians that had been trained to associate a specific substrate texture with the presence of food, they found that planarians retained this conditioned behavior after decapitation and head regeneration.

The behavior had been stored, apparently, not in the brain but elsewhere in the organism, possibly in the body wall or the nervous system peripheral to the cerebral ganglia. When a new brain was grown, it was informed by signals from the body and came to express behaviors that the original brain had encoded.

The implications for our understanding of memory, identity, and the relationship between the brain and the body are significant. But for this chapter's argument, the planarian experiment raises a question that will run through the entire book: if the information necessary for a new organization can survive the dissolution of the structure that originally encoded it, then the dissolution was not the end of the information. It was a transformation of the medium through which the information was expressed.

A Design Principle, Not a Specialization

The survey in this chapter is necessarily incomplete. Space does not permit a full treatment of metamorphosis in cnidarians, in tunicates, in polychaete worms, in crustaceans with their ecdysis-based growth cycles, or in the remarkable variety of plant developmental strategies that parallel metamorphic logic. What the examples do permit is a generalization.

Metamorphosis, understood as the capacity to reorganize radically rather than merely grow or change incrementally, is not the invention of any single evolutionary lineage. It appears in bacteria, in marine invertebrates, in insects, in amphibians, and in flatworms. It appears in organisms with sophisticated nervous systems and in organisms with none at all. It appears as a scheduled developmental event and as a response to environmental stress. It appears as a one-time lifetime event and as a capacity exercised repeatedly.

This breadth of distribution suggests something important about the nature of the capacity. When evolution finds the same solution independently across radically different lineages and conditions, it is usually because the solution represents a fundamental response to a fundamental challenge. The challenge here is this: how does a complex organized system change radically, rather than incrementally, when incremental change is no longer sufficient?

The answer that evolution has found, repeatedly and across the full diversity of life, involves the same basic sequence: accumulate the resources and the latent potential for a new organization while maintaining the current one; at the appropriate moment, release the current organization and allow its dissolution to provide the material and energy for the new one; let the latent elements express themselves and self-organize into the new form.

This is the design principle of metamorphosis. And the fact that it has been discovered independently by organisms ranging from soil bacteria to Monarch butterflies suggests that it is not a biological invention but a biological discovery, the discovery of a strategy that works because it is consonant with something true about how complex systems can change.

What that something is will be the subject of the next chapter.

III

The Metamorphic Genome

What genetics reveals about the blueprint of becoming

In January 1995, the Royal Swedish Academy of Sciences announced that the Nobel Prize in Physiology or Medicine would be shared among three scientists: Edward Lewis of Caltech, Christiane Nüsslein-Volhard of the Max Planck Institute for Developmental Biology, and Eric Wieschaus of Princeton University. The prize recognized their discoveries concerning the genetic control of early embryonic development in the fruit fly Drosophila melanogaster.

The work that won the prize had begun in the 1970s and involved a systematic screen of the Drosophila genome for mutations that disrupted the normal body plan of the developing fly. Nüsslein-Volhard and Wieschaus identified approximately one hundred genes whose disruption produced characteristic defects in fly development, and classified these genes into functional categories based on the nature of the defects they caused. Lewis, working independently over a longer period, had characterized the homeotic genes, a set of master regulatory sequences whose mutation produced the most spectacular developmental abnormalities: flies with legs where antennae should be, flies with four wings instead of two, flies with duplicated body segments.

These mutations were not merely curiosities. They revealed something fundamental about how bodies are built: that the identity of body parts, and the instructions for building them, are encoded in regulatory genes that act as master switches, turning on and off entire developmental programs. The homeotic genes of

Drosophila were the first clearly identified examples of what are now called Hox genes: a family of transcription factors whose DNA-binding domains are so highly conserved across animal species that equivalent genes, performing equivalent functions, are found in organisms as different as nematode worms, sea urchins, fish, and humans.

The Language of the Genome

The discovery of Hox genes was the first clear demonstration that the evolution of dramatically different body plans need not require the invention of entirely new genes. The genetic toolkit available to a fly and the genetic toolkit available to a vertebrate overlap substantially, with many of the same regulatory genes present in both and performing analogous developmental functions. Evolution, it turns out, works largely by changing where, when, and how much particular genes are expressed, rather than by continually inventing new genes from scratch.

This insight, which forms the core of the field now known as evolutionary developmental biology, or Evo-Devo, transformed our understanding of how metamorphosis is controlled and how it evolved.

The metamorphosis of holometabolous insects is controlled by a hormonal cascade that begins with a neuropeptide called prothoracicotropic hormone, or PTTH, secreted by cells in the brain. PTTH stimulates the prothoracic gland to produce ecdysteroids, a class of steroid hormones of which the most important is 20-hydroxyecdysone, commonly called 20E. Ecdysteroid signaling drives the molting process and, in coordination with juvenile hormone produced by the corpora allata, determines the identity of each successive molt: high juvenile hormone concentrations during ecdysteroid release produce larval molts; the absence of juvenile hormone during ecdysteroid release produces the pupal molt, initiating metamorphosis.

The downstream effects of this hormonal cascade are mediated through transcription factors, including members of the nuclear receptor superfamily, that activate or repress entire suites of target genes. These target genes, in turn, control the processes of histolysis, imaginal disc differentiation, and adult tissue morphogenesis that together constitute metamorphosis. The number of genes whose expression changes during metamorphosis is large: transcriptomic studies of developing Drosophila pupae have identified expression changes in thousands of genes across the pupal period.

What this genomic complexity reveals is that metamorphosis is not a simple developmental switch but an extraordinarily coordinated genetic program: a temporal sequence of gene expression changes that unfolds with high precision across days or weeks of pupal development, producing structures of remarkable complexity and functionality from undifferentiated precursor cells.

The Deep Conservation of Metamorphic Genetics

One of the most striking findings of comparative genomics is how deeply conserved the genetic machinery of developmental transformation is across the animal kingdom. The Pax6 gene, which in Drosophila is required for the development of the compound eye, is functionally equivalent to the Pax6 gene in vertebrates that is required for vertebrate eye development. When the mouse Pax6 gene is expressed in Drosophila embryos in contexts where eye development would normally be directed by the fly Pax6, it produces fly eyes, not mouse eyes. The gene specifies eye identity, and its identity-specifying function has been conserved across five hundred million years of evolution.

Similar deep conservation is found in the genetic regulation of limb development, heart development, and neural development across species that last shared a common ancestor hundreds of millions of years ago. This conservation has a profound implication: the genetic language in which developmental programs are written is ancient, and the vocabulary of that language, the set of regulatory genes and their interactions, is shared across an enormous range of animal species.

For the biology of metamorphosis specifically, this deep conservation suggests that the capacity for radical developmental reorganization is not a late evolutionary invention grafted onto an existing genetic architecture. The molecular components from which metamorphic programs are built, the nuclear receptors that respond to ecdysteroids, the transcription factors that activate imaginal disc differentiation, the cell signaling pathways that coordinate tissue dissolution and regrowth, all have homologs and evolutionary relatives in organisms that do not undergo complete metamorphosis. Holometabolism evolved by co-opting and recombining existing molecular machinery in a new temporal sequence, not by creating new genes from nothing.

This matters because it suggests that the potential for metamorphic transformation may be more widely distributed across the animal kingdom than the current distribution of holometabolous species would suggest. The genetic components are present. What varies is whether, and when, and under what conditions, they are assembled into a metamorphic program.

Epigenetics and the Environmental Trigger

The hormonal cascade that initiates insect metamorphosis does not operate in a vacuum. The decision to initiate metamorphosis is sensitive to environmental signals, including temperature, photoperiod, nutritional state, and population density. These signals are transduced through the neuroendocrine system, which modulates PTTH secretion and thereby adjusts the timing of metamorphic initiation to match environmental conditions.

This environmental sensitivity is mediated, in part, through epigenetic mechanisms: modifications to the structure of chromatin, the DNA-protein complex that packages the genome, that alter gene expression without changing the underlying DNA sequence. Epigenetic modifications can be stable across cell divisions, allowing them to propagate developmental memory from one cell generation to the next. And in some organisms, certain epigenetic modifications can be transmitted across generations, allowing the developmental experience of parents to influence the developmental trajectory of offspring.

The epigenetic dimension of metamorphosis suggests that the boundary between genetic programming and environmental influence in the control of transformation is not sharp. The genome encodes the possibility of metamorphosis and the rough outline of its program, but the precise timing, rate, and nature of the transformation are continually adjusted in response to environmental conditions. Metamorphosis is not purely determined from within; it is also responsive to what is happening outside.

This responsiveness will be relevant when we discuss, in later chapters, the environmental conditions that trigger or block metamorphic events in human psychology and in civilizational history. The general principle, that the timing and character of transformation are modulated by environmental context, is not unique to biology.

CRISPR and the New Power to Rewrite Transformation

In 2012, Jennifer Doudna of the University of California Berkeley and Emmanuelle Charpentier, then at the University of Vienna, published a paper in Science describing a mechanism derived from the bacterial immune system, CRISPR-Cas9, that could be adapted as a precise tool for editing genomic DNA. The paper launched a scientific revolution in genetic technology that within a decade had produced clinical applications, agricultural innovations, and research capabilities that had been inconceivable ten years earlier.

For Doudna and Charpentier, the discovery earned the 2020 Nobel Prize in Chemistry. For the biology of metamorphosis, CRISPR opened possibilities that are only beginning to be explored.

The most direct application is the ability to precisely edit the genes that control metamorphic timing and progression. Researchers have used CRISPR to knock out juvenile hormone signaling genes in Bombyx mori, the silkworm, producing larvae that bypass the pupal stage entirely and attempt to develop adult structures directly, with predictably disorganized results. They have used it to modify the expression patterns of homeotic genes in Tribolium castaneum, the flour beetle, producing insects with altered segment identities. They have used it to investigate the function of specific imaginal disc genes by disrupting their expression and observing the developmental consequences.

Beyond research applications, CRISPR-based tools are being developed for practical purposes that involve the manipulation of insect metamorphosis: gene drives designed to suppress mosquito populations by targeting the genes that control female fertility and metamorphosis; modifications to silk moth metamorphosis to improve the quality of commercial silk; and alterations to the metamorphic programs of agricultural pests to reduce their reproductive success.

These applications raise questions about the ethics and ecology of redesigning metamorphic programs, questions that will be addressed in Chapter Fourteen. For now, the significance of CRISPR for this chapter's argument is conceptual: the emergence of precise genomic editing tools has revealed, with new clarity, the degree to which metamorphosis is a genetic program that can be understood at the level of specific gene functions and interactions. What was previously a black box, the molecular machinery of transformation, has become a system that can be read, modified, and, within limits, redesigned.

Is Transformation Pre-Written?

The account of metamorphic genetics in this chapter supports a view that might be described as developmental predetermination: the butterfly is present, in some sense, in the caterpillar from the moment of the egg, encoded in the imaginal discs and in the hormonal and genetic programs that will eventually activate them. The transformation is not improvised. It is, in broad outline, pre-written in the genome, awaiting only the appropriate environmental and developmental conditions to unfold.

This view raises an obvious philosophical question: if transformation is pre-written in the genome, does the caterpillar have any choice about whether to transform? And by extension, when we observe metamorphic-like transformation in human beings, in cultures, in civilizations, is the outcome similarly pre-written in some way we do not yet fully understand?

The biological answer to the first question is nuanced. The capacity for metamorphosis is pre-written in the caterpillar's genome. The general trajectory of the transformation, a butterfly rather than a moth, a large butterfly rather than a small one, a particular species with its characteristic wing pattern, is broadly encoded. But the precise timing, the quality of the outcome, the survival of the individual through the vulnerable pupal phase, none of these are guaranteed or pre-written. The pre-written program provides the blueprint; the outcome depends on how the program unfolds in a particular environment under particular conditions.

The philosophical extension of this point will occupy us for much of the rest of this book. For now, it is enough to note that genetics has revealed something important: the capacity for radical transformation is encoded within living systems from their earliest developmental stages. It does not arrive from outside. It is not imposed by circumstance, though circumstances may be required to activate it. It is, in the deepest sense, internal: a potential present from the beginning, waiting for the conditions that will allow it to express itself.

What those conditions look like at the cosmic scale is where the next section of this book takes us.

IV

Stars Are Creatures Too

Stellar metamorphosis and the origins of transformation

The numbers are difficult to hold in mind, so let us begin with just a few of them.

The observable universe contains an estimated two trillion galaxies. The Milky Way, a mid-sized barred spiral galaxy, contains somewhere between one hundred billion and four hundred billion stars. The Sun, one of those stars, is four point six billion years old. It will live for approximately another five billion years before it expands into a red giant, swelling to engulf the inner planets, and then collapses into a white dwarf roughly the size of Earth.

At any given moment, somewhere in the observable universe, a massive star is dying. The death of a star with a mass greater than eight times the Sun's mass takes the form of a core-collapse supernova, one of the most energetic events in the universe: in the approximately ten seconds of its core collapse, a single supernova releases more energy than the Sun will emit over its entire ten-billion-year lifetime. The visible explosion that follows can outshine an entire galaxy for weeks.

These numbers describe the death of a star. What they do not capture is that the death is not an ending. It is, in a structural sense, the most metamorphic event in the universe.

The Life of a Star

A star begins as a nebula: a diffuse cloud of hydrogen and helium gas, typically containing also trace amounts of heavier elements produced in previous stellar generations. When a region of this gas cloud reaches sufficient density, through gravitational attraction or the shock wave of a nearby supernova, it begins to collapse under its own gravity. As it collapses, it heats. When the core temperature reaches approximately ten million degrees Celsius, nuclear fusion ignites: protons begin fusing into helium nuclei, and the enormous energy released generates outward radiation pressure that balances the inward pull of gravity. The star reaches a stable equilibrium: a main-sequence star.

This equilibrium is the star's larval phase, if we are using the metamorphic analogy. It is a period of sustained, relatively stable activity in which the star's primary function is energy production through fusion. The Sun has been in this larval phase for four point six billion years and will remain in it for approximately five billion more.

The end of the main-sequence phase comes when the star's core hydrogen is exhausted. The core, no longer producing energy through fusion, contracts under gravity and heats. The increased temperature ignites hydrogen fusion in a shell surrounding the now-inert helium core, causing the star's outer layers to expand dramatically. The star becomes a red giant or, for more massive stars, a red supergiant: still recognizably a star, but a star undergoing the beginning of a transition that will alter it fundamentally.

For stars like the Sun, the red giant phase ends with the gentle ejection of the outer layers into a planetary nebula and the cooling and contraction of the inert core into a white dwarf, a dense object roughly the mass of the Sun compressed into the volume of Earth. For massive stars, the story is more violent.

The Core Collapse: Nature's Most Violent Chrysalis

A star with a mass greater than roughly eight solar masses will, after exhausting hydrogen in its core, proceed through a series of successive fusion stages: helium fusion, then carbon fusion, then neon, oxygen, and finally silicon. Each successive fusion stage produces an increasingly heavy core element and proceeds on an increasingly short timescale. The silicon-burning phase, which produces iron, takes only a few days.

Iron is the end of the line for nuclear fusion in a stellar core. Iron has the highest binding energy per nucleon of any element: fusing iron consumes energy rather than producing it. When the iron core of a massive star reaches a critical mass, approximately one point four solar masses, electron degeneracy pressure can no longer support it against gravitational collapse. The core collapses.

The collapse takes approximately a quarter of a second. In that quarter second, a volume of material greater than the mass of the Sun is compressed from a diameter of several thousand kilometers to a diameter of approximately twenty kilometers, reaching densities of ten to the seventeenth power kilograms per cubic meter, comparable to the density of atomic nuclei. Protons and electrons are forced together to form neutrons, releasing enormous numbers of neutrinos. The collapsing material rebounds off the newly formed neutron core, generating a shock wave that propagates outward through the infalling stellar envelope.

The shock wave stalls briefly, revived by neutrino energy deposition, and then blows the outer layers of the star into space at velocities of five thousand to thirty thousand kilometers per second. This is the supernova explosion. The peak luminosity of a type II supernova is approximately ten billion times that of the Sun. The explosion produces, in a few seconds, quantities of heavy elements, including iron, nickel, gold, platinum, and uranium, that the star spent millions of years building in its core.

What remains at the center is the neutron star, or, if the collapsing mass is sufficient, a black hole. The neutron star spins rapidly, sometimes hundreds of times per second, and emits beams of radiation: a pulsar. The Crab

Pulsar, the remnant of a supernova observed by Chinese astronomers on July 4, 1054 CE, rotates thirty-three times per second and has been slowing at a measurable rate, its rotational energy gradually being radiated away.

You Are the Ash of Ancient Metamorphoses

The significance of stellar metamorphosis for this book's argument is not primarily aesthetic, though the spectacle of a supernova is genuinely extraordinary. It is chemical.

The Big Bang produced, almost exclusively, hydrogen and helium. Lithium was produced in small quantities; everything heavier was not. The atoms of every element heavier than helium that exists in the universe today were produced either in the cores of stars or in the explosive violence of supernova events.

The carbon in your body was produced in the core of a red giant star through the triple-alpha process, in which three helium nuclei fuse to form a carbon nucleus. The oxygen you breathe was produced primarily through the fusion of carbon and helium. The iron in your blood was produced in the cores of massive stars and scattered into the galaxy by supernova explosions. The calcium in your bones, the nitrogen in your DNA, the phosphorus in your ATP, all were produced in stars and distributed through the galaxy by the metamorphic events of stellar death.

The astrophysicist Lawrence Krauss put it plainly in his 2012 book: every atom in your body came from a star that exploded. The sun provides the energy that drives the chemistry of life, but it did not produce the atoms. Those atoms were produced in stars that lived and died before the Sun was formed, and they were distributed into the interstellar medium by the metamorphic events through which those stars ended one form of existence and began another.

In this sense, your body is not merely made of stardust in the poetic sense that phrase is often used. It is made of the specific products of specific metamorphic events: stellar deaths that converted one set of organized structures into another, releasing in the process the material that would eventually, billions of years later, be incorporated into a planet, into life, into you.

The caterpillar-to-butterfly metaphor for this process has obvious limits. A star has no nervous system, no subjective experience, no agency. The metamorphosis of a star is purely physical. But the structural parallel with biological metamorphosis is not merely aesthetic: in both cases, a stable organized system exhausts the conditions that sustained its original structure; undergoes a phase of dissolution or radical reorganization; and produces, through that reorganization, elements and structures that could not have been produced under the original conditions. The new universe is richer in heavy elements because old stars died.

Cosmic Recycling and the Question of Continuity

One of the most striking features of stellar metamorphosis is the degree to which the products of stellar death become the raw material of stellar birth. Supernova remnants expand into the interstellar medium, enriching the galactic gas clouds from which new stars form with the heavy elements produced in the explosion. New stars that form from this enriched material inherit the chemical complexity accumulated across multiple stellar generations.

The Sun is a third-generation or later star: the heavy elements in its composition and in the planets that formed from the solar nebula are the products of multiple previous stellar generations, each of which lived, evolved, and died, contributing its heavy elements to the galactic pool. Without those preceding stellar metamorphoses, the Sun would not have the chemical composition to host a planet capable of producing carbon-based life. We exist in part because previous stars died in ways that released the elements necessary for our existence.

This is the cosmic version of the pattern identified in Chapter One. The dissolution of one organization provides the material for a new organization with greater complexity and capability. The caterpillar's dissolved tissues become the building material for the butterfly's wings. The exploded star's scattered elements become the building material for new solar systems, planets, and, eventually, living creatures. The pattern scales.

Whether this scaling reflects something deep about the physics of complex systems, or whether it is a coincidence of analogy that should not be pressed too hard, is a question that will occupy us in Part Two and Part Four. For now, it is enough to notice that the pattern exists at the cosmic scale, that it is not metaphorical but physical, and that its existence at this scale suggests that whatever is being observed is not a biological quirk but a feature of how organized systems, of any kind, can undergo radical change.

V

Quantum Metamorphosis

Transformation at the foundations of physical reality

In 1801, the English physician and polymath Thomas Young performed what would later be considered one of the most important experiments in the history of physics. He directed a beam of light through a barrier with two narrow parallel slits and observed the pattern produced on a screen behind the barrier. If light were composed of particles, as Isaac Newton had argued, the two slits should produce two bands of light on the screen. What Young observed instead was a pattern of alternating bright and dark bands: an interference pattern of the kind produced by waves passing through two openings and interacting with each other on the other side.

Young's double-slit experiment established that light behaves as a wave. A century later, the same experiment was performed with individual electrons, particles of matter with a definite mass and charge. The electrons also produced an interference pattern, suggesting that they, too, behave as waves. And here the experiment became philosophically interesting in a way that simple wave-particle duality does not capture.

When the experiment was modified to include a detector at the slits, to register which slit each electron passed through, the interference pattern disappeared. The electrons began behaving as particles rather than waves. When the detection was turned off, the interference pattern returned. The act of measuring which path the electron took seemed to change the nature of the electron's behavior.

This is not a paradox produced by some limitation of the measurement apparatus. It is not an artifact of the experimental design that more refined instruments would resolve. It is, as far as seven decades of quantum mechanical investigation have been able to determine, a fundamental feature of physical reality at the quantum scale: the properties of a quantum system are not fully determined until they are measured. Before measurement, the system exists in a superposition of all possible states. The act of measurement collapses this superposition into a single definite state.

The relevance of this to a book about metamorphosis is not that quantum mechanics is a metaphor for transformation, though the metaphorical resonances are real. It is that quantum mechanics describes a layer of physical reality in which the concept of identity, the notion that something is what it is independently of observation and interaction, does not straightforwardly apply.

Superposition and the Potentiality of Being

The quantum state of a system, described mathematically by the wave function, represents not what the system is but what it might be: a distribution of probabilities over all possible measurement outcomes. An electron in a superposition of spin-up and spin-down states is not secretly either spin-up or spin-down, with the observer merely ignorant of which. According to the standard interpretation of quantum mechanics, it is genuinely both, until a measurement forces it into one state or the other.

This is philosophically radical in a way that deserves emphasis. The classical picture of reality, which most of us carry unreflectively, is that things have definite properties that exist independently of observation. A coin tossed in the air is either heads or tails; we just do not know which until we look. Quantum mechanics challenges this picture at the deepest level: at the quantum scale, things may genuinely not have definite properties until they are observed, where observation means some form of physical interaction that collapses the superposition.

The relevance to metamorphosis is conceptual rather than mechanistic. What quantum mechanics reveals is that the distinction between potential and actual, between what a system might become and what it currently is, is not merely epistemic but physical. Potentiality is real. The unrealized states of a quantum system are not merely hypothetical; they are physically present and causally relevant, as the interference pattern in the double-slit experiment demonstrates. The wave function, which describes the superposition of all possible states, is not just a description of our ignorance but a description of something real about the physical system.

In a biological organism, the imaginal cells carry the potential for the adult form while the caterpillar persists. In a quantum system, the unrealized eigenstates of the wave function persist alongside the realized state. In both cases, the future organization is present within the current system in a form that is real but not yet expressed. The transition to expression, whether through the hormonal cascade of pupation or through the interaction of measurement, involves a fundamental change in the organization of the system.

Phase Transitions: Metamorphosis at the Molecular Scale

Quantum mechanics is not the only domain of physics in which metamorphic-like transformations occur. Phase transitions, the transformations of matter between different physical states, offer another and more macroscopic example of radical reorganization.

When water at atmospheric pressure is cooled to exactly zero degrees Celsius, it transitions from liquid to ice. The transition is not gradual: at one degree below zero, water is liquid; at one degree below zero minus the smallest measurable temperature increment, the water is ice. The transition happens at a precise temperature and involves a fundamental reorganization of the molecular structure: in liquid water, molecules are held together by hydrogen bonds in a disordered, constantly shifting network; in ice, they are arranged in a regular, crystalline lattice.

The transition between these two phases involves a substantial change in physical properties: density, thermal conductivity, mechanical rigidity, optical properties. The same collection of water molecules, with the same chemical composition and the same internal energy relationships, organizes itself into dramatically different macroscopic structures depending on temperature. The difference between liquid water and ice is not a difference in composition but a difference in organization, produced by a change in the thermodynamic conditions.

Physicists characterize phase transitions by their order. A first-order phase transition, like the melting of ice, involves an abrupt change in some property of the system, in this case density, at a specific transition point, and requires the absorption or release of latent heat. A second-order phase transition, like the transition of iron from ferromagnetic to paramagnetic behavior at the Curie temperature of 770 degrees Celsius, involves a more gradual change in properties and no latent heat. Both involve a fundamental reorganization of the system's structure.

The language of phase transitions has been applied by physicists and complexity theorists to systems far beyond simple matter. Network science has identified phase transitions in the connectivity of random graphs: below a certain threshold edge density, a random graph consists of many small disconnected components; above the threshold, a giant connected component suddenly emerges, containing most of the nodes. The transition is abrupt and represents a fundamental change in the global organization of the system produced by a local change in one parameter.

The concept of a phase transition, a qualitative reorganization of a system's structure triggered by quantitative change in a control parameter, will appear again in Chapter Ten, when we discuss the dynamics of civilizational transformation, and in Chapter Sixteen, when we examine the conditions under which collective human metamorphosis might occur.

Entanglement and the Non-Local Coherence of Transformation

Quantum entanglement is the phenomenon in which two quantum systems become correlated in such a way that the quantum state of each cannot be described independently of the other, even when the systems are separated by arbitrarily large distances. When a measurement is performed on one entangled system and collapses its superposition, the state of the other system is simultaneously determined, regardless of the distance between them.

Einstein famously described this as spooky action at a distance and argued, in a 1935 paper co-authored with Boris Podolsky and Nathan Rosen, that it implied quantum mechanics was an incomplete description of physical reality. Subsequent theoretical work by John Bell and experimental tests by Alain Aspect and collaborators in 1982 established that the correlations of entanglement cannot be explained by any local hidden variable theory: the non-local correlation is real, not an artifact of pre-existing correlations established when the systems last interacted.

For this book's purposes, entanglement is interesting not as a mechanism for metamorphosis but as a demonstration of non-local coherence: the idea that a transformation in one part of a system can be instantaneously correlated with a transformation in another part, even when the two parts are widely separated. The unity of the transformation is not mediated by any signal traveling through space; it is a property of the entangled quantum state.

Whether anything analogous to quantum entanglement occurs in biological or social metamorphic systems is an open question that this book will not try to settle. The point is more modest: physics has discovered, at its most fundamental level, that the assumption of independent local properties, the assumption that everything is what it is in its own location and that changes here are not correlated with changes there, is not generally valid. Reality at the quantum scale is more holistic than classical intuition allows.

The Reality of Process

The picture that emerges from quantum mechanics, and from the physics of phase transitions, is of a physical reality that is fundamentally processual rather than thing-like. The classical picture of the universe as a collection of objects with fixed properties, undergoing changes that are always in principle reversible and always mediated by local interactions, does not survive examination at the quantum scale.

What survives is a picture of reality as a web of processes and interactions, in which properties are relational and contextual rather than intrinsic and fixed. A particle does not have a definite position or momentum until it interacts with something. A phase of matter is a stable pattern of organization under certain conditions that may give way to a different pattern under different conditions. A quantum system in superposition is not secretly one thing or another; it is genuinely a superposition, and the transition to a definite state is produced by an interaction that transforms the system.

This is the ontology that the philosopher Alfred North Whitehead was groping toward in his process philosophy, developed in the 1920s and 1930s, in which the fundamental constituents of reality are not substances with properties but events and processes of becoming. Whitehead's metaphysics was developed before the full implications of quantum mechanics were understood, and the resonance between his process ontology and the quantum mechanical picture of reality has been noted by several philosophers of physics.

The relevance to metamorphosis is this: if reality at the physical level is fundamentally processual, then change, including radical change, is not an anomaly or an exception in the story of physical reality. It is the story. The relative stability that we perceive as the persistence of things is a stability of pattern, of organized process, rather than the persistence of some underlying unchanging substance. Things are what they are because of the processes that maintain them. When those processes change, the things change. When they change radically, the things change radically.

Metamorphosis, understood in this light, is not an unusual event that requires special explanation. It is what happens when the organizing processes of a system reach a point at which they can no longer maintain the current pattern and must reorganize into a new one. The question is not why metamorphosis occurs, but why it sometimes does not.

VI

The Dark Chrysalis

What the invisible universe teaches us about transformation

Consider what we cannot see.

The atoms that make up everything you can directly observe, every star, every planet, every gas cloud, every living thing, constitute approximately five percent of the total energy content of the observable universe. The remaining ninety-five percent is composed of two things that have never been directly detected: dark matter, which makes up approximately twenty-seven percent of the total, and dark energy, which constitutes approximately sixty-eight percent. These figures come from the European Space Agency's Planck satellite mission, whose 2018 data release provides the most precise measurement of the universe's composition yet achieved.

We know dark matter exists because of its gravitational effects. We do not know what it is.

We know dark energy exists because the expansion of the universe is accelerating, which requires some form of energy distributed through empty space. We do not know what it is either.

The universe is therefore, in a very precise sense, mostly hidden from us. The part we can see and measure is a small minority of what is actually there. And the part we cannot see is not merely an absence of the visible. It is a positive presence with real physical effects, organizing the behavior of the visible universe in ways that we can observe but cannot directly investigate.

This has a structural parallel to the chrysalis that is not merely figurative.

Fritz Zwicky and Vera Rubin: The Discovery of the Invisible

The first serious proposal that the universe contains matter beyond what can be observed came from the Swiss astronomer Fritz Zwicky, working at Caltech in 1933. Zwicky was studying the Coma Cluster, a large cluster of galaxies approximately three hundred twenty million light-years from Earth, and found a discrepancy between the mass of the cluster inferred from the velocities of its member galaxies and the mass that could be accounted for by the observable matter. The galaxies were moving too fast: the visible mass was insufficient to explain their velocities. Zwicky proposed that the cluster must contain substantial amounts of unseen matter, which he called dunkle Materie, dark matter, to account for the discrepancy.

The proposal was largely ignored for four decades. The evidence that finally compelled the astronomical community to take dark matter seriously came from Vera Rubin of the Carnegie Institution of Washington, working with her colleague W. Kent Ford. Beginning in the late 1960s and continuing through the 1970s, Rubin and Ford measured the rotation curves of spiral galaxies: the relationship between the orbital velocity of stars and their distance from the galactic center.

In a system where most of the mass is concentrated at the center, like a solar system where most of the mass is in the sun, orbital velocities should decrease with distance from the center, following Kepler's laws. Rubin and Ford found instead that the rotation curves of spiral galaxies are essentially flat: stars at the outer edges of galaxies orbit at approximately the same velocity as stars closer to the center. This flatness implies the presence of a large amount of mass distributed throughout the galaxy and extending well beyond its visible disk, a dark matter halo invisible to optical and radio telescopes but exerting gravitational effects on the visible matter within it.

Rubin's work, published across a series of papers in the 1970s and 1980s, established dark matter as a mainstream scientific concept. Subsequent observations using gravitational lensing, the bending of light by mass, have provided independent confirmation that large amounts of unseen matter are distributed throughout the universe, organized into halos and filaments that form a cosmic web, the large-scale structure on which galaxies and clusters of galaxies are threaded like beads.

The Scaffolding That Cannot Be Seen

The cosmic web of dark matter is not passive background. It is the structural scaffold on which the visible universe is organized. Computer simulations of cosmic structure formation, run without dark matter, do not produce the observed distribution of galaxies. The filaments, walls, and voids of the large-scale structure of the universe, the structures that contain clusters and superclusters of galaxies, emerge in simulations only when dark matter is included in the initial conditions.

Dark matter halos around galaxies are not simply the gravitational containers that hold the visible stars. They shape the growth of galaxies by determining which regions of space attract enough matter to form stars and which do not. The visible galaxy is, in a structural sense, a consequence of the invisible dark matter halo: the shape, size, and rotation of the galaxy you see are determined substantially by the shape and density profile of the dark matter you cannot see.

This inversion of the visible and the structurally primary is the aspect of dark matter most relevant to the metamorphic argument. We tend to assume that what we can see is what is most real, or at least most causally important. The dark matter evidence suggests otherwise. The invisible structures are the ones that most fundamentally determine the organization of the visible.

The scaffolding is more primary than the building it supports.

The Unconscious as Dark Matter

The parallel to human psychology is imprecise but worth drawing. The concept of the unconscious, as developed in the psychoanalytic tradition and subsequently investigated through cognitive science and neuroscience, describes a domain of mental activity that influences behavior and experience without being directly accessible to conscious awareness. In the Freudian formulation, the unconscious contains repressed material that shapes behavior through indirect pathways. In the broader cognitive science formulation, the unconscious encompasses the vast majority of information processing that underlies conscious experience but is not itself conscious.

Research in social psychology and cognitive neuroscience over the past thirty years has repeatedly demonstrated that much of what determines human behavior and decision-making is not accessible to conscious report. Studies of priming, implicit memory, and automatic processing have shown that cognitive processes initiated without conscious awareness can influence judgments, preferences, and behaviors in systematic and powerful ways. The work of Daniel Kahneman, summarized in his 2011 book Thinking, Fast and Slow, provides an accessible account of the degree to which fast, automatic, unconscious processes, what Kahneman calls System 1, dominate human cognition and behavior.

In the context of personal metamorphosis, the invisible dimension of the psyche plays a role structurally analogous to dark matter: it constitutes the primary organizing force for transformation, shaping the direction and character of change through processes that the conscious mind does not directly observe or control. The dreams, the somatic experiences, the recurring patterns of attraction and aversion, the inexplicable senses of rightness or wrongness about particular choices: these are the visible signs of invisible processes that are doing the actual work of psychological reorganization.

This does not mean that conscious intention is irrelevant to transformation. It means that conscious intention is one input into a much larger and largely invisible system of organizing processes. The person who believes their transformation is entirely under conscious control, driven purely by rational decision and deliberate effort, is probably wrong, in the same way that an astronomer who believes the galaxy's structure is entirely produced by the visible stars is wrong.

Emergence and the Hard Problem

The final theme of this chapter is emergence: the phenomenon in which properties appear at one level of organization that were not predictable from, or present in, the components at a lower level. Water is wet; individual water molecules are not wet. Flocking birds exhibit coordinated motion; individual birds do not have a representation of the flock's trajectory. The brain produces consciousness; individual neurons do not appear to be conscious.

Emergence is a well-established phenomenon in physics, chemistry, and biology. The existence of emergent properties does not require anything supernatural or outside the causal order; it requires only that the behavior of complex systems cannot be fully predicted from the behavior of their components, because the interactions between components at scale produce properties that were not present at the component level.

In the context of metamorphosis, emergence is the mechanism by which the new organization produces capacities that the old organization did not have. The butterfly's capacity for flight is not present in any individual cell of the caterpillar. It emerges from the specific organization of the butterfly's body, including the wing structure, the flight muscles, the aerodynamic properties of the wing surface, and the neural circuits that control wing movement. All of these are present in the caterpillar in potential, encoded in the imaginal discs, but the capacity for flight is not present as such. It emerges only when the new organization is assembled.

The hard problem of consciousness, the problem of explaining why physical processes give rise to subjective experience at all, is in many ways the most extreme case of emergence: why is there something it is like to be a brain, when there is presumably nothing it is like to be a neuron? The philosopher David Chalmers, who coined the phrase the hard problem in a 1994 paper, has argued that the emergence of consciousness from neural processes is not explicable by the same functional and informational analysis that explains other emergent properties, and that consciousness may require a fundamental expansion of our physical ontology to accommodate.

Whether Chalmers is right about consciousness is a question that continues to be debated with intensity in philosophy of mind and cognitive science. For this chapter's purposes, the relevance of the hard problem is its illustration of a general point: the visible, the measurable, the functionally analyzable, may not be all there is. The most significant dimension of a metamorphic event may be the one that is least directly observable from outside.

The chrysalis, from outside, is a dead thing. The supernova, for the first several months of its explosion, is an expanding sphere of hot gas, undifferentiated and apparently chaotic. The grieving person, in the depth of their dissolution, appears to those around them to be simply in pain. In all three cases, something is happening that the external view does not reveal and cannot fully describe. The most active, most important, most generative processes are the invisible ones.

This is what the dark chrysalis teaches: that the absence of visible progress is not evidence of the absence of transformation. It may be evidence of its presence.

VII

The Brain That Remakes Itself

Neuroplasticity and the neural chrysalis

On December 10, 1996, a thirty-seven-year-old neuroanatomist named Jill Bolte Taylor woke at 5:30 in the morning with a sharp pain behind her left eye. Over the next four hours, she observed her own brain suffering a massive hemorrhagic stroke, her scientific training allowing her to recognize, with increasing clarity, each capacity as it was lost. First went the ability to distinguish the boundaries of her body from the boundaries of the room. Then language. Then sequential thinking. Then the memory of who she was.

What she describes, in her 2008 memoir My Stroke of Insight, is not merely the deterioration of function but something more ambiguous: a parallel state that emerged as the left hemisphere's activity declined. The right hemisphere's experience, freed from the left's categorical organization, produced a sense of oceanic unity, of being without boundary, that she describes with unmistakable intensity as a profound and unexpected experience of peace.

Taylor survived because her brother called emergency services in time. She spent the next eight years in recovery, relearning how to walk, how to talk, how to read, how to do mathematics. She has spoken publicly about the recovery process as a kind of choosing, a decision made thousands of times over those years about which neural patterns to encourage and which to let go. Some of what she rebuilt was the same as what she had had before the stroke. Some of it was different. The woman who emerged from the recovery was a neuroanatomist who could speak and write and think clearly, and who was also, unmistakably, not quite the same person who had entered the stroke.

Taylor's case is extreme. But the neuroscience it illustrates applies, in less dramatic form, to every human brain at every stage of life.

The Discovery of Adult Neuroplasticity

For most of the twentieth century, the dominant view in neuroscience held that the adult brain was essentially fixed in structure. Neurons could die and their connections could weaken, but the fundamental architecture of the adult nervous system was, after a critical period in early childhood, no longer subject to significant change. Santiago Ramón y Cajal, the founding figure of modern neuroscience, wrote in 1913 that in the adult centres the nerve paths are something fixed, ended, immutable. Everything may die, nothing may be regenerated.

This view was not entirely wrong. The rate of structural change in the adult brain is, under normal conditions, substantially lower than in the developing brain. But it was wrong in important respects, and the accumulation of evidence to the contrary over the second half of the twentieth century amounted to one of the most significant revisions in the history of neuroscience.

The researcher who did most to establish adult neuroplasticity as a mainstream scientific concept was Michael Merzenich of the University of California San Francisco, who began publishing evidence for cortical reorganization in the late 1960s and continued developing and defending the case through the following four decades. Merzenich's early work showed that the somatosensory cortex of adult monkeys reorganized after peripheral nerve damage: the cortical territory representing an amputated finger was taken over by the representations of adjacent fingers, demonstrating that the topographic map of the body on the cortical surface was not fixed but could be redrawn in response to changed inputs.

The implication was direct: if the cortical map could reorganize in response to the loss of input, it could presumably also reorganize in response to new or intensified input. This was confirmed in studies of musicians and other practitioners of intensive, long-term sensorimotor skill. String musicians show expanded cortical representation of the left hand, which performs the more complex and demanding fingerwork. Braille readers show expanded cortical representation of the reading fingertip. Taxi drivers who must navigate large and complex cities develop enlarged hippocampal volumes.

The London Taxi Driver Study

The taxi driver study is worth describing in detail because it is one of the most widely cited and clearly interpretable demonstrations of experience-dependent structural change in the adult human brain. Eleanor Maguire and colleagues at University College London recruited sixteen licensed London taxi drivers and compared their brains, using structural MRI, to fifty non-taxi-driver controls matched for age, sex, and education. The study was published in the Proceedings of the National Academy of Sciences in 2000.

The hippocampus, a bilateral structure in the medial temporal lobe, is known to be important for spatial navigation and the encoding of episodic memories. Maguire's team found that the London taxi drivers had significantly larger posterior hippocampal volumes than the controls, and that the enlargement was positively correlated with the amount of time the driver had spent as a taxi driver: drivers with more years of experience had larger posterior hippocampi. A follow-up study confirmed that this difference was not pre-existing, a selection effect in which individuals with larger hippocampi were more likely to become taxi drivers, but was acquired through the demands of the job.

To qualify as a licensed taxi driver in London, an individual must pass a test called The Knowledge, which requires the memorization of twenty-five thousand streets within a six-mile radius of Charing Cross, along with the routes to and from major landmarks. The test typically requires three to four years of study and multiple test attempts to pass. The hippocampal enlargement observed in licensed drivers is, in all probability, the structural consequence of this intensive, prolonged spatial memory training.

The human brain rewrites its own architecture in response to sustained experience. This is not metaphorical. It is the growth of new synaptic connections, the strengthening of existing ones, and, in the case of hippocampal neurogenesis, the actual generation of new neurons in the adult brain, a phenomenon once considered impossible that was definitively established in adult human tissue by Fred Gage of the Salk Institute in 1998. The adult brain is capable of structural metamorphosis. What it requires is the appropriate stimulus and the time and conditions to reorganize.

Trauma and Post-Traumatic Growth

Trauma is the closest thing in human experience to the hostile conditions that trigger bacterial sporulation or the environmental stress that initiates insect metamorphosis ahead of schedule. It is an overwhelming event or series of events that exceeds the capacity of the existing psychological organization to process and integrate. The immediate response to trauma is typically the disruption of normal functioning: disrupted sleep, hyperarousal, intrusive memories, emotional dysregulation, and, in severe cases, the structural changes in the brain associated with post-traumatic stress disorder, including reduced hippocampal volume, hyperreactivity of the amygdala, and altered prefrontal cortex function.

The neurobiological profile of PTSD is, in metamorphic terms, the profile of a system that has entered a dissolution phase without completing the transition to a new organization. The old organization has been disrupted, the normal processing of experience has been overwhelmed, but the resources and conditions for rebuilding have not yet been established.

But there is another outcome of trauma that receives substantially less attention in the clinical literature and in public discourse: post-traumatic growth. The term was coined by Richard Tedeschi and Lawrence Calhoun at the University of North Carolina Charlotte in a 1996 paper that proposed a framework for understanding the positive psychological changes that some individuals report following severely challenging life circumstances. The changes identified in their research include increased personal strength, new possibilities, relating to others, appreciation of life, and spiritual change.

The prevalence of post-traumatic growth is contested, partly because measuring positive psychological change following trauma is methodologically challenging. But the phenomenon is consistently reported. A review published in Clinical Psychology Review in 2016 found that between 30 and 70 percent of survivors of a wide range of traumatic events, including cancer diagnosis, bereavement, military combat, natural disaster, and sexual assault, reported positive changes in some domains of their lives following the event. The range of reported prevalence reflects the heterogeneity of trauma, of measurement approaches, and of populations studied.

Post-traumatic growth is not universal, and its presence does not negate the reality or the severity of the suffering that preceded it. It is not a silver lining that justifies or minimizes the trauma. It is, rather, evidence that the human brain and psyche, like the biological systems described in earlier chapters, sometimes respond to conditions that overwhelm their current organization by reorganizing into something new rather than simply restoring what was there before.

Meditation and the Intentional Chrysalis

If trauma is the involuntary version of neural metamorphosis, meditation may be the closest thing to a voluntary one. The scientific study of the effects of sustained meditation practice on brain structure and function has grown substantially over the past two decades, producing findings that, while sometimes overstated in popular accounts, are sufficiently robust to merit serious attention.

Sara Lazar of Massachusetts General Hospital and Harvard Medical School published a study in 2005 in NeuroReport demonstrating that experienced meditators, with an average of nine years of practice, showed significantly greater cortical thickness in the prefrontal cortex, the right anterior insula, and the sensory cortices compared to matched non-meditators. The insula, a cortical region associated with interoception, the perception of the body's internal state, showed the most pronounced thickness differences, particularly in older participants for whom age-related cortical thinning would normally be expected.

Subsequent studies have found a range of structural and functional differences between long-term meditators and non-meditators, including greater gray matter density in the hippocampus and posterior cingulate cortex, reduced gray matter density in the amygdala, and altered patterns of connectivity in resting-state brain networks. The default mode network, a set of brain regions associated with self-referential thought, mind-wandering, and rumination, shows reduced activity in experienced meditators during rest and during meditation, a finding associated with reduced tendencies toward anxious or depressive rumination.

What meditation appears to do, at the neural level, is to alter the default patterns of information processing in the brain: shifting resources away from the automatic, reactive, self-referential processes and toward more deliberate, present-centered, and interoceptively aware modes of experience. These are not changes in the content of experience but changes in the architecture of the system that produces experience.

In metamorphic terms, meditation is a practice of voluntary dissolution: a sustained, regular practice of releasing habitual patterns of thought and self-identification, not in order to destroy the self but in order to allow it to reorganize around different organizing principles. The meditator is not simply relaxing, though relaxation may be a component. They are, over the course of years of practice, performing a slow, intentional metamorphosis of the processes through which they experience and respond to reality.

Your brain is not fixed. Every significant experience, every sustained practice, every encounter with conditions that exceed your current capacity: these are all invitations to neural metamorphosis. Most of them pass unaccepted. Some do not. And the ones that do not are the ones that, in retrospect, feel most like becoming.

VIII

Grief, Loss, and the Cocoon You Did Not Choose

Suffering as metamorphic process

You know the morning. You have known it, or you will. It is the morning after something ends: a relationship, a marriage, a career, a vision of your own future, a person you loved. You wake with the immediate, physical weight of it, before conscious thought has fully assembled. The loss is in the body before it is in the mind. You stay in the bed longer than you should. The ordinary world outside the window looks the same as it always did, and that sameness is somehow the most offensive thing about it.

You are inside a chrysalis. You do not know this yet, because no one has told you, and because from the inside a chrysalis is not experienced as a protective structure or a developmental phase. It is experienced as being trapped in something that is dissolving you.

This chapter is about that experience. But it is also, ultimately, about what the experience means when it is placed in the larger frame that this book has been building. Not to minimize it or explain it away. Not to offer premature consolation. But to make it, as far as possible, legible.

The Body of Grief

Grief is not, despite a great deal of cultural messaging to the contrary, primarily a mental or emotional experience. It is a somatic one. The physiological response to significant loss involves the same neural systems engaged by physical pain: imaging studies have shown that the anterior cingulate cortex and the anterior insula, brain regions that process the affective dimension of physical pain, are activated during social rejection and loss. The phrase that a loss is felt in the body is not a metaphor. It is neuroscience.

The physiological changes that accompany acute grief are extensive. Cortisol levels rise, suppressing immune function and altering glucose metabolism. Sleep architecture is disrupted, reducing the time spent in slow-wave and REM sleep, the stages most associated with emotional regulation and memory consolidation. Cardiovascular stress increases, significantly enough that the phenomenon of fatal cardiac events following bereavement has been documented in the medical literature: a review published in JAMA Internal Medicine in 2014 found that the risk of a heart attack was twenty-one times higher in the twenty-four hours following the death of a significant person than at a baseline comparison period.

This is what it means when people say that someone died of a broken heart. In some cases, they literally did. The body processes loss as an emergency.

Why should this be? From an evolutionary perspective, the intensity of the grief response appears disproportionate to its adaptive value. Prolonged grief impairs the griever's ability to function, to seek food, to defend territory, to maintain social bonds. What is the adaptive logic of a response that is so physiologically costly?

One answer, proposed by several evolutionary psychologists, is that grief is the price of attachment: a biological system that evolved to enforce the maintenance of close social bonds, by making their loss physiologically aversive, will produce strong attachment bonds that benefit the organism under normal conditions. Grief is the system working as designed, registering the loss of something that the body has been organized around.

In metamorphic terms: grief is histolysis. It is the dissolution of the organizational patterns, the neurological, physiological, and psychological patterns that constituted the self-that-was-attached. Those patterns do not dissolve painlessly, just as the caterpillar's muscles do not dissolve painlessly. The pain is the cost of the dissolution, and the dissolution is the cost of becoming.

Kübler-Ross and the Stages Reconsidered

Elisabeth Kübler-Ross's model of the stages of grief, first published in her 1969 book On Death and Dying, is one of the most widely known and widely misunderstood frameworks in the history of psychology. The five stages she identified, denial, anger, bargaining, depression, and acceptance, were not, in Kübler-Ross's original formulation, a fixed sequence through which all grieving people necessarily pass in order. They were a descriptive framework drawn from her conversations with terminally ill patients, intended to give clinicians and caregivers a vocabulary for understanding the psychological states they would encounter.

The model was subsequently popularized and simplified in ways that Kübler-Ross herself later expressed ambivalence about. The stages became, in popular understanding, a prescription rather than a description: a sequence that grieving people were supposed to move through in order, with acceptance as the correct endpoint. People who did not experience all five stages, or who experienced them in a different order, or who oscillated back and forth between them, sometimes concluded that they were grieving incorrectly, a belief with no empirical or clinical foundation.

The research of George Bonanno of Columbia University, published across a series of papers and summarized in his 2009 book The Other Side of Sadness, complicates the stages model further. Bonanno's longitudinal studies of bereaved individuals found that the most common trajectory following significant loss was not the extended period of debilitating grief followed by gradual recovery that the stage model implies, but a pattern of resilience: a relatively brief period of disruption followed by a return to near-baseline functioning. This resilient response was found in approximately sixty-five percent of bereaved individuals in his studies.

The minority of bereaved individuals who experienced prolonged and severe grief showed what Bonanno identified as a distinct pattern, later classified in the diagnostic literature as Prolonged Grief Disorder: persistent, intense yearning for the deceased; intrusive thoughts about the loss; difficulty accepting the death; and significant impairment of social, occupational, or other functioning beyond twelve months following the loss. The prevalence of this pattern in bereaved populations is estimated at approximately ten to fifteen percent.

In 2019, David Kessler, a collaborator of Kübler-Ross's in the late years of her career, proposed a sixth stage to add to the original five: meaning. Drawing on his own experience of grief following his son's death and on decades of clinical work with the bereaved, Kessler argued that the movement beyond acceptance toward the active creation of meaning from a loss represents a distinct and important phase in the grief process. Not the consolation of finding a silver lining, but the harder and more dignified work of asking what the loss reveals, what it requires, and what new organization of a life might be built in its aftermath.

Liminality: The Anthropology of the In-Between

In 1909, the French ethnographer Arnold van Gennep published Les Rites de Passage, a comparative study of rituals that mark transitions between social states: birth, puberty, marriage, death. Van Gennep identified a three-phase structure in these rituals across cultures: separation from the previous state, a liminal or threshold period of ambiguity and transition, and reincorporation into a new state. The liminal phase, from the Latin limen meaning threshold, is the period in which the individual is no longer what they were and not yet what they will become.

The anthropologist Victor Turner, working in the 1960s and 1970s, developed Van Gennep's concept of liminality into one of the most productive analytical frameworks in cultural anthropology. Turner observed that liminal individuals, those in the threshold phase of a ritual transition, are characterized by ambiguity and paradox: they have shed their previous social identity but not yet acquired a new one; they are simultaneously powerful and vulnerable; they are often associated with disorder, creativity, and the dissolution of normal social categories.

The liminal phase, Turner argued, is not merely a passage between two stable states. It is a distinctive mode of existence with its own properties, its own logic, and its own social and psychological functions. The disorientation and vulnerability of the liminal phase serve important purposes: they loosen the individual's attachment to their former identity and social position, creating the psychological flexibility that allows a new identity to be taken on. The dissolution is not accidental. It is the mechanism.

Turner's framework was developed in the context of ritual transitions that are socially scripted and publicly marked. But the structure he identified applies with equal force to involuntary transitions: divorce, job loss, serious illness, bereavement. These transitions also move through a liminal phase, and the liminal phase has the same properties that Turner identified in ritual contexts: ambiguity, vulnerability, dissolution of previous identity markers, and the potential for creative reorganization.

One of the most psychologically damaging features of contemporary Western culture's treatment of grief and loss is its impatience with the liminal phase. The cultural messages around loss, in their various forms, communicate that the dissolution phase should be completed quickly, that prolonged liminality is pathological, that the appropriate response to loss is active effort to return to previous functioning as rapidly as possible. This cultural impatience with the chrysalis phase is, in metamorphic terms, exactly wrong.

The Intelligence of Dissolution

Grief dissolves things that needed to be dissolved. Not because the things dissolved were worthless or mistaken, but because the self that was organized around them has been permanently disrupted by the loss, and a new organization is required. The dissolution is not a mistake. It is the first phase of an adaptive process.

This claim requires careful handling. It is not an argument that loss is good, or that grief should be welcomed, or that the things we lose are better lost. It is an argument about what grief does biologically and psychologically when it is allowed to proceed through its natural course rather than being suppressed or abbreviated. What grief does is reorganize the self around the absence of what has been lost. This reorganization takes time, and it requires the kind of deep internal process that looks from outside like doing nothing, like being stuck, like failing to move on.

The person in the chrysalis phase of grief is not stuck. They are in the most metabolically active phase of their transformation. The reorganization is occurring. The imaginal cells of the new self, the latent capacities, values, and relationships that the previous organization could not accommodate, are beginning to activate. The resistance from the existing system, the pull toward the previous identity and the previous attachments, is real and painful. But it is the immune response of a system that has not yet completed its transformation, not evidence that the transformation is going wrong.

You are not broken. You are mid-dissolve. It is not pleasant to be mid-dissolve. The caterpillar, if it were capable of experience, would presumably not enjoy it either. But the dissolution is not the story's end. It is the story's mechanism.

What emerges may surprise you. Not because the surprise is guaranteed, not because every chrysalis produces a butterfly, because some do not. But because what is being built from the dissolved material of your previous organization is not simply a reconstruction of what was there before. It is something new, assembled from the same elements according to a blueprint that the dissolution has activated.

The chrysalis you did not choose is doing exactly what chrysalises do.

IX

Who Are You After?

Identity, continuity, and the philosophy of becoming

Suppose that in the year 2050, teleportation technology has become available. You enter a teleporter in New York, which scans every atom of your body at its exact position and quantum state, transmits this information at the speed of light to a receiver in London, and constructs, from new matter, an exact replica of you in London. Your body in New York is destroyed in the scanning process. The replica wakes in London with all your memories, your personality, your characteristic ways of moving and speaking and thinking. It believes it is you. Is it?

This thought experiment, used by the philosopher Derek Parfit in his landmark 1984 work Reasons and Persons, is designed to elicit intuitions about personal identity and to expose their inconsistency. Most people, when asked, say that they would not willingly undergo this process, that the person who wakes in London is not them but a copy of them, and that their own experience would end in the scanner in New York. Yet the replica has everything that ordinarily makes a person who they are: their memories, their personality, their psychological continuity with their previous self.

Parfit's deeper argument is that this reluctance to endorse the teleportation case exposes a confusion in our ordinary concept of personal identity: that we believe personal identity to be a determinate fact that must be either fully preserved or fully lost, when in reality it is a matter of degree, consisting in overlapping chains of psychological connectedness that can be more or less continuous, more or less dense, more or less similar to the identity relations we find in ordinary life.

The puzzle Parfit poses is abstract. But it becomes concrete the moment you try to answer the question that titles this chapter: after a profound personal transformation, who are you?

The Ship of Theseus

The Ship of Theseus is one of the oldest paradoxes in Western philosophy. The ship of the hero Theseus, according to the Athenian tradition reported by Plutarch, was preserved in the harbor of Athens as a sacred memorial. As its planks rotted, they were replaced one by one, until eventually every plank of the original ship had been replaced. Was the resulting ship still the Ship of Theseus?

Thomas Hobbes added a further complication: suppose someone had collected all the original planks as they were replaced, repaired them, and assembled them back into a ship. Now two ships exist: the continuously maintained one in the harbor, and the reassembled one built from the original materials. Which is the Ship of Theseus?

The paradox has direct application to the human body. The cells of most human tissues are replaced over time through normal cell division and turnover. Red blood cells survive approximately four months. The cells lining the intestine are replaced every two to four days. Bone cells turn over on a timescale of roughly ten years. The neurons of the cerebral cortex are among the longest-lived cells in the body, and most persist for a human lifetime, but their molecular components, the proteins and lipids and other molecules that constitute them, are continuously replaced through normal cellular metabolism. At the molecular level, you are a largely different physical system than you were ten years ago.

You nonetheless feel continuous with your past self. You remember being the person who was ten years younger, even though most of the molecules that constituted that person are gone. This sense of continuity is real and psychologically important. But what is its basis, if not material continuity?

The Buddhist Answer and Its Modern Resonance

The Buddhist concept of anatta, often translated as no-self or non-self, denies the existence of a permanent, unchanging self that persists through time and experience. In the Buddhist analysis, what we call the self is a flowing process of physical and mental events, conventionally designated by a name and a body but without any underlying essence that is the bearer of those events. The Buddha, in the Anattalakkhana Sutta traditionally dated to the early period of his teaching, argues that none of the five aggregates, form, feeling, perception, mental formations, and consciousness, is the self, because each is impermanent and subject to change.

The practical implication of anatta is not nihilism but freedom: if there is no fixed self that must be defended and maintained at all costs, then the dissolution of a previous identity in the course of transformation is not a catastrophe but a natural event, requiring no more grief than the replacement of old cells by new ones. The Buddhist practitioner is, in this sense, cultivating metamorphic literacy: the capacity to observe the arising and passing of mental and physical events without clinging to any particular configuration as definitive.

The resonance between the Buddhist analysis of selfhood and contemporary psychological research on narrative identity is striking. Dan McAdams of Northwestern University has proposed, in work developed over three decades, that personal identity in adulthood is constituted primarily by the life story that a person constructs about themselves: an ongoing narrative that integrates past, present, and anticipated future into a coherent, meaningful account of who one is and how one came to be that way. Identity is not a thing but a story, and like all stories, it can be revised.

McAdams's research suggests that what distinguishes individuals who successfully navigate major life transitions from those who do not is, in part, the quality of the narrative revision they are able to perform: the ability to incorporate the disruption of the transition into a coherent story that acknowledges what was lost, accounts for the change, and projects a plausible future self. This is, in metamorphic terms, the narrative equivalent of imaginal disc activation: the capacity to construct a new story from the materials of the old one, not by erasing the old but by integrating it into a larger account.

Mourning Your Previous Selves

One of the most neglected dimensions of personal transformation is the grief that accompanies the loss of previous versions of oneself. When a person undergoes a major transformation, they do not only gain something. They lose something. The person they were before the transformation, the caterpillar-self, is gone in a real and significant sense. The beliefs, the habits, the relationships, the ways of being in the world that characterized that previous self are no longer available in the same way, if they are available at all.

This loss is legitimate. It deserves acknowledgment. The tendency in the culture of personal development to celebrate transformation without acknowledging the loss that accompanies it is a distortion that leaves people confused about why transformation, even when it is going well, is so often accompanied by grief.

You mourn the person you were not because that person was better or more valuable than who you are becoming, but because that person was you, and they are no longer here. The relationships that person had, the competencies they possessed, the particular quality of experience that characterized their way of being: these are genuinely gone, and their going is a loss even when what replaces them is an improvement.

Winnicott, the British psychoanalyst and pediatrician, wrote about the developmental necessity of what he called the destruction of the object: the process by which a child, and by extension any maturing psyche, must be able to engage with the world in ways that could potentially damage or destroy it, and to discover that the world survives this engagement. What a person discovers in that survival, Winnicott argued, is the reality and independence of the external world, as distinct from an internal fantasy. The mourning of previous selves has something of this character: you engage the reality of the loss fully enough to discover that you survive it.

Continuity Through Transformation

The question who are you after? does not have a clean answer. After major transformation, you are neither simply the same person as before nor simply a different person. You are the same genetic organism, with a continuous history, some continuous relationships, and some degree of psychological continuity with your previous self. You are also significantly different in values, behavior, beliefs, emotional patterns, and relational capacities.

What this means, practically, is that the question itself may be wrong. The question presupposes that identity is a binary: either preserved or lost. The metamorphic evidence, at every scale from the caterpillar to the civilization, suggests that identity through transformation is not binary but gradient: it is a matter of what is preserved, what is lost, and what is entirely new, and the relationship between these three categories is what makes each transformation unique.

The more productive question is not who are you after, with its implicit demand for a definitive answer, but how do you relate to what you were? What of your previous self do you carry forward as genuine continuity? What do you lay down with genuine grief? What do you recognize as the beginning of something you had not been before?

These are not philosophical abstractions. They are the practical questions of a metamorphic life, and learning to ask them well is part of what it means to be metamorphically literate. You are, at any moment, in some relationship with the several selves that have preceded you and the self that is emerging. Managing that relationship with honesty and dignity, neither clinging to what has been lost nor repudiating it, is the ongoing work of a transformative life.

X

Civilizations in the Chrysalis

How societies dissolve and what determines whether they emerge

On September 4, 476 CE, a Germanic chieftain named Odoacer deposed Romulus Augustulus, the last emperor of the Western Roman Empire, and sent the imperial regalia to Constantinople with a message that the West no longer needed a separate emperor. The date is used by historians as the conventional endpoint of the Western Roman Empire, though the reality of Rome's dissolution was considerably more complex, protracted, and geographically uneven than the date implies.

For the people living through the fifth century in the western provinces, the experience was not of a transition between historical periods, legible in retrospect as the boundary between ancient and medieval. It was of progressive deterioration: roads unmaintained, trade routes disrupted, populations declining, urban centers shrinking, literacy rates falling. The elaborate administrative, military, and economic system that had organized the lives of tens of millions of people across three continents was dissolving, and it was not obvious to those living through the dissolution that anything would replace it.

From the perspective of six hundred years later, the outline of what replaced it was clear: the medieval synthesis of Christian theology, Germanic customary law, and residual Roman administrative structure that formed the basis of European civilization for a millennium. But the medieval world did not emerge immediately from the fall of Rome. It emerged slowly, haltingly, from the resources that survived the dissolution, over a period of several centuries during which the continuities and discontinuities with Roman civilization were worked out in ways that no one planned or predicted.

The Fall of Rome, seen through the metamorphic lens, is not primarily a story of collapse. It is a story of dissolution and reorganization, and understanding it as such changes the questions we ask about it and the lessons we draw from it.

What Survives Dissolution

The historians Peter Heather, Bryan Ward-Perkins, and others have documented the real and serious costs of the Western Empire's dissolution. Archaeological evidence shows sharp declines in trade volumes, pottery quality, building construction, and long-distance economic integration in the fifth and sixth centuries. Population estimates for the western provinces suggest a decline from approximately seventy million in the second century CE to roughly forty-five million by the seventh century. The costs were real.

But the dissolution was not total. Several things survived from the Roman organizational system and became the building blocks of what emerged. The Latin language survived, in modified forms, as the basis of the Romance languages and as the international language of the Church and later of medieval scholarship. Roman law survived, transmitted through the Corpus Juris Civilis of Justinian, codified in Constantinople in 529 to 534 CE, and through local applications of Roman legal principles in the western kingdoms. The Christian church, which had grown within the Roman imperial structure, survived the political dissolution of that structure and provided an institutional continuity across the discontinuity of political power.

These are the imaginal cells of the civilizational metamorphosis: the structures that survived the dissolution of the old organization and provided the building materials for the new one. They did not produce the medieval world by a simple process of continuation; they were transformed in the process of being incorporated into a new synthesis. But their survival was what made the medieval synthesis possible rather than simply a collapse into undifferentiated barbarism.

The Scientific Revolution as Imaginal Disc Activation

In February 1543, the Polish astronomer Nicolaus Copernicus sent the final manuscript of De Revolutionibus Orbium Coelestium to his publisher in Nuremberg. He died, according to legend, on the day the first printed copy arrived at his bedside in Frauenburg, on May 24 of that year. The book proposed that the Earth revolves around the Sun rather than the Sun around the Earth, the heliocentric model of the solar system, and in doing so initiated the most consequential intellectual transformation in the history of Western civilization.

The Copernican revolution is often described as a revolution in astronomy. It was that, but it was much more. It displaced humanity from the center of the cosmos, and with that displacement, it began undermining the entire framework of medieval scholastic cosmology in which the structure of the physical universe was aligned with and reflective of the structure of Christian theology. The heavens were no longer a system of perfect crystalline spheres rotating around the Earth in a hierarchy that mirrored divine order. They were a physical system operating according to mathematical laws that could be discovered by observation and reason.

This shift took a century and a half to work out its full implications, through Tycho Brahe's precise observations, Galileo's telescopic discoveries and his confrontation with the Inquisition, Kepler's laws of planetary motion, and finally Newton's Principia Mathematica of 1687, which provided the mathematical framework within which the heliocentric system made complete physical sense. The Scientific Revolution was not a sudden event but a sustained metamorphic transition of European intellectual culture, from a framework organized around theological authority and classical text to one organized around empirical observation and mathematical description.

The imaginal cells of the Scientific Revolution were present within medieval scholasticism in the form of the empiricist strand of natural philosophy represented by Roger Bacon and others who argued, against the dominant Aristotelian tradition, that knowledge of the natural world must be based on direct observation. They were present in the mathematical tradition of Islamic scholarship transmitted to Europe through translation in the twelfth and thirteenth centuries. They were present in the commercial culture of Italian city-states, which required precise calculation and provided the social support for the kind of individual inquiry that the Scientific Revolution demanded.

The dissolution phase of the transition, the period in which the medieval framework was being dismantled faster than the new one was being assembled, was experienced as deeply disorienting. The phrase the world turned upside down, which circulated widely in seventeenth-century England to describe the social and intellectual upheaval of the period, captures the phenomenology of civilizational chrysalis: the old orientation is gone; the new one is not yet established; the in-between is characterized by anxiety, violence, and the feeling that nothing is certain.

Five Conditions for Civilizational Metamorphosis

The comparative history of civilizational transformations suggests a set of conditions that appear to influence whether a period of civilizational dissolution produces genuine metamorphosis, a new synthesis that is more complex and more capable than what preceded it, or collapse, in which the dissolution is not followed by a new organization but by a sustained reduction in complexity and capability.

The first condition is the survival of imaginal cells: the institutions, practices, and knowledge systems that carry the blueprint for a new organization must survive the dissolution phase. When a civilization's dissolution destroys its literacy, its legal traditions, its scientific knowledge, or its institutional memory, the materials for a new synthesis are not available, and recovery is slower and more limited.

The second condition is sufficient dissolution. A civilization that resists dissolution too effectively, that maintains its existing organizational structures through suppression and force in the face of conditions that have made those structures obsolete, tends to produce brittle failure rather than metamorphosis. The dissolution must be deep enough to release the resources and the organizational space that the new synthesis requires.

The third condition is connection among imaginal elements. The surviving elements of a previous civilization must be able to connect with each other and with new elements emerging from the changed conditions. The medieval synthesis was possible because the Church provided an institutional structure within which the surviving elements of Roman culture, including Latin literacy, legal tradition, and philosophical thought, could be connected with the new Germanic social structures and the theological framework that gave the synthesis its organizing principle.

The fourth condition is time. Civilizational metamorphosis does not happen quickly. The transition from the Roman world to the medieval synthesis took several centuries of uncertain, painful, contested process. Attempts to accelerate the transition, to force the emergence of a new synthesis before the dissolution has done its necessary work, tend to produce unstable structures that do not last.

The fifth condition is something harder to name: a kind of collective willingness to let go of the previous organization's most valued features while preserving its most essential ones. This is the civilizational equivalent of the caterpillar's immune system eventually standing down in the face of the imaginal cells. It requires a degree of collective psychological flexibility that is rare and difficult and never guaranteed.

Whether the civilization currently underway is in a position to meet these five conditions is the subject of the book's final chapter. For now, it is enough to note that the conditions have been met before, in different ways and with different outcomes, and that the historical record gives us something more useful than either optimism or pessimism: a set of factors to examine carefully as we try to understand where we are.

XI

Is Change Real?

The philosophy of transformation from Heraclitus to the present

Fragment 91. That is the number assigned by scholars to the surviving text in which Heraclitus of Ephesus, writing around 500 BCE, made the statement that has become the most quoted fragment of pre-Socratic philosophy: You cannot step twice into the same river. The river Heraclitus had in mind was, most probably, the Cayster or the Maeander near Ephesus on the Aegean coast of what is now Turkey. But the river he was describing was something larger: the nature of reality itself.

What made the fragment scandalous, at least to those of Heraclitus's contemporaries who understood it, was not the observation about rivers, which is banal enough. It was the implication: if the river into which you stepped a moment ago no longer exists, and the you that stepped into it no longer quite exists either, then the persistence of identity through change is a kind of useful fiction, a cognitive convenience that we impose on a reality that is, in its actual nature, a continuous flowing process with no fixed points.

Heraclitus did not conclude from this that nothing was real. He concluded, rather, that what is real is not the fixed things we ordinarily take to be real but the process itself, the logos, the rational principle of transformation, the law by which opposites give way to each other and all things flow. The river is real. The water that constitutes it at any given moment is a temporary configuration. The logos that keeps the river being a river through all the changes in its water is the most real thing about it.

This is not a bad description of what a butterfly is. The butterfly is real. The caterpillar cells and adult cells that constitute it at different moments of its life cycle are temporary configurations. The developmental program that keeps the organism being the organism it is through the radical transformations of its body is the most real thing about it. Heraclitus, without knowing anything about holometabolism, had identified the philosophical structure of metamorphosis.

Parmenides and the Case Against Change

Parmenides of Elea, writing a generation after Heraclitus and explicitly in response to him, reached the opposite conclusion. In his poem The Way of Truth, preserved in fragments through later commentators, Parmenides argued that genuine change is logically impossible. For something to change, it must come to be something that it was not. But something that is not does not exist. Therefore, for something to become what it is not, it must move from existence to non-existence, which is impossible.

The argument is more sophisticated than this summary suggests, and Parmenides drew from it the startling conclusion that reality is a single, undivided, unchanging being: the One, as it became known in later discussions. The appearance of change and plurality that we experience through the senses is, on Parmenides's account, an illusion or a lesser form of truth, to be distinguished from the genuine, rational knowledge of the One that is unchanging.

The Parmenidean challenge to the reality of change is not simply a philosophical curiosity from antiquity. It reappears, in various forms, throughout the history of philosophy and science. The concept of conservation laws in physics, the principle that certain quantities such as energy, momentum, and electric charge are conserved through all physical processes, is in part a Parmenidean legacy: through all the apparent changes in physical reality, something fundamental is preserved. The question of what that something is, and whether its preservation is what justifies calling the changed system the same system, is a question that physics does not fully answer.

The Platonic tradition, which drew on Parmenides, introduced the concept of eternal forms or ideas, unchanging archetypes of which temporal things are imperfect copies. The changing world of experience is real but less real than the unchanging world of forms. This metaphysical hierarchy has had enormous influence on Western culture's evaluation of change: the tendency to treat the permanent as more real and more valuable than the transient, and to regard transformation with suspicion as a movement away from the ideal.

The butterfly's wings do not last. The organism that produced them is, evolutionarily speaking, only as real as the genes it passes on to the next generation. But while it flies, it is more fully what it is than the caterpillar was. The metamorphic perspective challenges the Platonic hierarchy: perhaps the form that is achieved through transformation is not less real than an unchanging archetype but more real, because it is produced through a process that has tested and refined it. Perhaps becoming is not a fall from being but its fullest expression.

Hegel's Metamorphic Logic

Georg Wilhelm Friedrich Hegel's philosophical system, developed in works including the Phenomenology of Spirit of 1807 and the Science of Logic of 1812 to 1816, represents one of the most ambitious attempts in the history of philosophy to think rigorously about the nature of change and development. Hegel's dialectic, the famous thesis-antithesis-synthesis structure, is often described in simplified and distorted forms in secondary literature, but in its original formulation it offers genuine insights into the structure of transformation.

For Hegel, the movement of thought, and of reality as he understood it, which was ultimately identical with thought, is not a linear accumulation of truths but a process in which each stage of development carries within itself a contradiction, a negation, that drives it beyond itself into a new stage. The new stage does not simply replace the old one: it is what Hegel calls Aufhebung, a German word that simultaneously means cancellation, preservation, and elevation. The old stage is cancelled in its immediate form but preserved in the new form that has superseded it.

This is, structurally, a precise description of metamorphosis. The caterpillar is cancelled in its immediate form, the larval body is dissolved and reorganized. But it is not simply destroyed. Its genetic identity is preserved. Some of its neural information is preserved. The metabolic resources accumulated by the larval body are preserved and used in the construction of the adult. And the adult is, in the Hegelian sense, an elevation: a form that encompasses what the caterpillar was while transcending its limitations.

The resonance between Hegel's dialectic and the biological logic of metamorphosis has not, to my knowledge, been noted in the philosophical literature. But it is more than coincidental. Both describe a pattern in which a stable organization generates, through its own internal dynamics, the conditions for its own dissolution and supersession by a more complex organization. Whether this pattern is, as Hegel believed, the fundamental logic of all development, or whether it is a common pattern in complex adaptive systems without being a universal metaphysical law, is a question that the rest of this chapter will address.

Whitehead and Process Philosophy

Alfred North Whitehead, the British mathematician and philosopher who spent the last decades of his life at Harvard, developed in Process and Reality of 1929 what remains the most sustained philosophical attempt to take seriously the claim that reality is fundamentally processual rather than thing-like. Whitehead's speculative cosmology, which he called the philosophy of organism, proposed that the fundamental constituents of reality are not substances with properties but events of experience, which he called actual occasions.

Each actual occasion, in Whitehead's system, is a process of becoming: it takes in data from its past, experiences and responds to that data in a process that has an aesthetic dimension, and completes itself in a moment of self-determination that then becomes data for subsequent occasions. Nothing in the universe is simply there, static and inert. Everything is an event, a process of becoming from what has been to what is being made.

Whitehead's metaphysics was developed independently of quantum mechanics but has repeatedly attracted the attention of philosophers of physics who find in it resources for thinking about the ontological implications of quantum theory. The process ontology fits naturally with the quantum mechanical picture of a reality in which properties are not pre-existing facts but are constituted through interactions and measurements.

For the argument of this book, Whitehead's significance is his insistence that process is not a secondary or derivative feature of reality but its primary character. Change, in a process metaphysics, is not something that happens to things. It is what things are. The stability that we perceive as the persistence of things is a stability of pattern within an ongoing process, not the persistence of some underlying unchanging substance. Metamorphosis, understood in this light, is not a disruption of the normal order of things. It is the normal order of things, made visible.

What the Debate Reveals

The philosophical debate about change, which has run continuously from Heraclitus and Parmenides to the present, reveals something important about human psychology: we are deeply ambivalent about transformation. The Parmenidean impulse to find something permanent and unchanging beneath or behind the flux of experience is not merely a logical error or a philosophical eccentricity. It reflects a genuine feature of human experience: we need some degree of continuity and stability to function. We need to be able to recognize our environment, our relationships, and ourselves from moment to moment. A reality that changed completely at every instant would be uninhabitable.

The Heraclitean impulse to insist on the reality of change is equally important and equally grounded in genuine experience: we know that things change, that we change, that the world of our childhood is not the world we inhabit as adults. A philosophy that denies the reality of this change in favor of an eternal unchanging realm is, whatever its logical merits, in tension with lived experience.

The metamorphic perspective offers a way of holding both impulses: recognizing the reality of change while identifying what persists through change, not as an unchanging substance but as a continuous developmental process. The butterfly is not the same as the caterpillar in the sense that matters most to the Parmenidean: it does not have the same form, the same structure, the same functional identity. But it is related to the caterpillar by a process of continuous transformation, and that relationship is real, meaningful, and in certain respects, the most important thing about both of them.

Reality is change. What persists through change is not a fixed substance but an ongoing process. Understanding this is not merely a philosophical achievement. It is a practical one. It changes how we live inside the changes that come to us, and how we participate in the changes we ourselves initiate.

XII

The Ethics of Becoming

What we owe our past selves and those we leave behind

A woman I know, whom I will call Elena, left a fifteen-year marriage in her early forties. She had entered the marriage as a certain kind of person and had, over the preceding three years, become, through a combination of therapy, spiritual practice, and the slow accumulation of self-knowledge, substantially a different kind of person. Her values had shifted. Her understanding of what she needed from intimate relationship had shifted. Her sense of what she was capable of and what she owed herself had shifted.

Her former husband had not undergone the same transformation. From his perspective, and from the perspective of many of their mutual friends, what Elena had done was change the rules in the middle of the game. She had made commitments as one person and then, in effect, declared them no longer binding by becoming another person. She had transformed herself out of a relationship that, from the outside, appeared to be functioning, and in doing so she had caused real, substantial harm to someone who had not consented to losing her.

Was Elena's transformation selfish? Did she owe something to her previous self's commitments? Does radical personal transformation carry ethical obligations that ordinary change does not?

These are not easy questions, and this chapter will not pretend otherwise. But they are questions that a book about metamorphosis is obligated to address.

The Nietzschean Model and Its Shadow

Friedrich Nietzsche's concept of self-overcoming, Selbstüberwindung in German, is the most celebrated philosophical celebration of personal transformation in the Western tradition. In Thus Spoke Zarathustra, published in four parts between 1883 and 1885, Nietzsche presents the figure of Zarathustra as an exemplar of the self-overcoming that he takes to be the highest human aspiration: the willingness to continually surpass one's current form, to refuse the comfort of fixed identity, to become what one is by constantly exceeding what one was.

The concept is exhilarating and not without genuine philosophical depth. The recognition that identity is not a given but an achievement, that the self is something created rather than discovered, that clinging to a previous version of oneself is a form of bad faith, these are insights with real force. Nietzsche was one of the first Western philosophers to take seriously the processual nature of selfhood that Buddhist philosophy had long recognized.

But the Nietzschean model has a shadow side that becomes visible when you examine the cases of individuals who have used the language of self-overcoming and transformation to justify a wholesale abandonment of prior commitments, relationships, and obligations. The person who leaves a marriage, a friendship, a community, or a set of responsibilities in the name of transformation, without acknowledging the real costs of the departure to those affected, is not performing self-overcoming in any philosophically defensible sense. They are performing escape, and using the vocabulary of growth to avoid the vocabulary of responsibility.

The distinction matters. Self-overcoming, in Nietzsche's own account, is not a painless process and is not a process in which one's obligations to others are simply dissolved by the act of transformation. It is a process of genuine struggle with what one is, including one's entanglements with others and one's commitments to them. The courage that self-overcoming requires is not primarily the courage to leave but the courage to face honestly what leaving costs and what it means.

The Ethics of Transformation in Recovery

The context in which questions about the ethics of personal transformation are most urgently and practically addressed is recovery from addiction. The twelve-step model of addiction recovery, developed by Alcoholics Anonymous beginning in 1935 and subsequently extended to other addictive behaviors, involves a process of radical personal transformation that explicitly addresses the ethical dimensions of that transformation.

The fourth through ninth steps of the twelve-step program are directly concerned with the ethics of becoming. The fourth step involves making a searching and fearless moral inventory of oneself: an honest accounting of the harms done to others in the course of the addictive behavior. The eighth and ninth steps involve making a list of all persons harmed and making direct amends to them wherever possible, except where doing so would injure them or others.

The logic of these steps is instructive. The program does not assume that transformation releases the transforming person from accountability for the actions of their former self. The person who emerges from addiction recovery is genuinely different from the person who was in active addiction, in many of the ways that matter. But they are also the same person in the way that matters ethically: they are continuous with the person who caused harm, they share that person's history, and they are therefore the appropriate party to acknowledge and, where possible, repair it.

The making of amends is not primarily a legal or psychological transaction. It is an acknowledgment that transformation does not erase the past. The new self inherits the history of the selves that preceded it, including the obligations those selves incurred. The caterpillar's debts do not dissolve with the caterpillar. The butterfly is answerable for them.

Betrayal Trauma and the Person Left Behind

The person on the other side of someone else's transformation deserves ethical attention that is often absent from discussions of personal growth. Jennifer Freyd of the University of Oregon developed the concept of betrayal trauma in a 1994 paper to describe a specific form of psychological injury that occurs when a trusted figure violates the trust of the person who depends on them. The key variable in betrayal trauma is not the severity of the harmful act but the degree of trust violation: harms committed by intimates cause a distinctive and particularly damaging form of psychological injury because they undermine the relational security on which the victim depends.

When a person undergoes radical personal transformation in ways that fundamentally alter or end their significant relationships, the people who loved and depended on the previous version of them may experience something with the character of betrayal trauma: the person they trusted has been replaced, from their perspective, by someone they do not fully recognize, who no longer offers the same relational security, and who may have made the transformation partly at their expense.

This experience is real and its pain is legitimate. The transforming person does not dissolve the experience by pointing to the authenticity or necessity of their transformation. The fact that a transformation was genuine, necessary, and ultimately beneficial for the person who underwent it does not, by itself, address the harm experienced by those affected by it.

The ethical obligation created by this situation is not to avoid transformation in order to protect others from the disruption it causes. That would be its own form of self-betrayal and would, in most cases, ultimately harm everyone involved. The obligation is to acknowledge the harm, to take it seriously rather than dismissing it as the other person's resistance to growth, and to offer whatever genuine repair is possible.

Growth Versus Escape: A Practical Distinction

One of the most practically important questions in the ethics of transformation is the distinction between genuine growth and flight. Both can look similar from the inside, particularly in the early stages. Both involve a departure from a previous organizational form. Both involve a period of increased psychological discomfort. Both can be described, sincerely or insincerely, in the language of personal development.

The distinction between them is not always accessible to the transforming person themselves, at least not immediately. One of the functions of good therapeutic work, good spiritual direction, and good friendship is to provide external perspective that helps a person distinguish between the two. But some markers of the distinction can be named.

Genuine growth tends to be characterized by an increase in the capacity for relational honesty, an increased willingness to acknowledge the complexity of one's situation rather than simplifying it, and a growing ability to hold both the necessity of one's transformation and the legitimate pain of those affected by it without needing to resolve the tension by dismissing either. Flight tends to be characterized by a simplification of the narrative, a reduction of the complexity of one's previous situation to something that justifies the departure, and an impatience with the other's perspective that is recognizable as the avoidance of genuine complexity.

Transform without accountability is not liberation. Genuine metamorphosis does not require the abandonment of moral seriousness. The butterfly that emerges from the chrysalis is not free of the fact that it was once a caterpillar. It carries that history in its genome, in its memory, in the cells that have persisted through the transformation. The lightness of the butterfly is not the lightness of having dropped a burden. It is the lightness of a body organized for a new mode of existence, while remaining, in all the ways that matter, accountable for having inhabited the old one.

XIII

Does Metamorphosis Have a Direction?

Teleology, complexity, and the question of meaning

Approximately 13.8 billion years ago, in the first fractions of a second following what we call the Big Bang, the universe contained nothing but an extremely hot, dense plasma of quarks, gluons, and leptons. There were no atoms. There were no molecules. There were no stars, no planets, no galaxies, no chemistry, no biology, no minds. There was only energy, and the forces that would eventually shape it.

Today, the same matter and energy that existed in that primordial plasma has organized itself into the structures we observe: two trillion galaxies, each containing hundreds of billions of stars, many of those stars orbited by planets, at least one of those planets hosting a biosphere containing an estimated eight million species, including one species that has developed language, mathematics, music, and the capacity to ask questions about its own origins and destiny.

The trajectory from quarks to questions is the most dramatic metamorphosis in the record of cosmic history. And it raises, with an urgency that no philosophical evasion can fully suppress, the question that this chapter confronts: does this trajectory have a direction? Is there something that the universe, or life, or the evolutionary process, is becoming? And if there is, does that imply meaning?

I want to approach these questions with as much honesty as I can, which means acknowledging from the outset that they touch the boundary between what science can establish and what philosophy and theology have traditionally addressed. I am not a theologian, and this book is not a work of theology. But I am also not willing to dismiss these questions as meaningless or to pretend that the evidence is neutral on them.

Aristotle's Entelechy and the Inner Directedness of Development

Aristotle introduced the concept of entelechy, from the Greek en, meaning in, telos, meaning end or purpose, and echo, meaning to have or hold, to describe the principle by which a thing contains within itself the form of what it is striving to become. The acorn has the oak as its entelechy: not as an external cause that pulls the acorn toward oak-hood from outside, but as an internal organizing principle that directs its development from within. The entelechy is not a soul in the supernatural sense but a natural principle of directed development.

The concept was dismissed for centuries as unscientific, a vestige of teleological thinking incompatible with the mechanistic worldview of modern science. The modern biological equivalent of entelechy, the developmental program encoded in the genome, is understood in mechanistic terms: the direction of development is produced by the interaction of genes and their products, not by a non-physical goal-directed principle.

But the dismissal of Aristotelian teleology may have been too hasty. The developmental program encoded in the genome is, in a real sense, directed: it guides the organism toward a specific end state, the adult form, rather than simply unfolding randomly. The end state, though it is specified by molecular mechanisms rather than by a non-physical principle, is nonetheless real and normative: development can succeed or fail, depending on whether the end state is reached. The genome contains an entelechy in the functional sense, even if not in Aristotle's metaphysical sense.

When we scale this observation from individual development to evolutionary history, the picture becomes more complex and contested. Does evolution, considered as a whole, have a direction? Does it tend toward something?

Complexity Theory and the Arrow of Time

The physicist and cosmologist Eric Chaisson of Tufts University has developed, over several decades, a quantitative measure of complexity that he calls energy rate density: the rate at which energy flows through a system per unit mass. This measure increases systematically across the history of the universe, from galaxies and stars, to planets and planetary ecosystems, to individual organisms, to human brains, to human technology.

The energy rate density of a typical galaxy is roughly ten times that of a star. A plant's energy rate density is roughly a thousand times that of a star. An animal's is roughly ten thousand times that of a star. The human brain, with its extraordinarily high metabolic rate for its mass, has an energy rate density approximately one hundred thousand times that of a star. Human technology, including the most energy-intensive computational systems, reaches energy rate densities several orders of magnitude higher still.

Chaisson's measure is not without critics, and its interpretation as evidence of directional evolution is contested. But it captures something that is genuinely puzzling: across the history of the universe, structures of increasing organizational complexity, able to process increasing amounts of energy per unit mass, have consistently emerged. This is not what a naive application of the second law of thermodynamics would predict. The second law states that entropy, a measure of disorder, increases in closed systems. The universe as a whole is moving toward greater disorder. Yet within it, ordered structures of increasing complexity continue to emerge.

The resolution of this apparent paradox is that living systems and other complex structures are open systems, not closed ones: they maintain and increase their internal order by exporting entropy to their surroundings. Life decreases local entropy by increasing entropy elsewhere, using the flow of energy through them as the resource for this order-maintenance. The question of whether this persistent tendency toward increasing organizational complexity in open systems represents a genuine arrow, a direction, in the history of the universe, or merely a statistical tendency produced by the particular initial conditions of the Big Bang, is one that neither physics nor biology has fully resolved.

Teilhard de Chardin and the Omega Point

Pierre Teilhard de Chardin, the French Jesuit priest and paleontologist who spent much of his career in China and died in New York in 1955, proposed in his posthumously published work The Phenomenon of Man a vision of cosmic evolution as a directed process moving toward an ultimate state of maximum complexity and consciousness that he called the Omega Point. Teilhard argued that evolution is not merely a biological process but a cosmic one, and that its direction, from matter to life to mind to spirit, is inscribed in the structure of the universe itself.

Teilhard's vision was received with hostility by both scientists and Catholic theologians: by scientists because it appeared to introduce teleological and theological concepts into evolutionary biology, and by theologians because it seemed to naturalize the spiritual in ways that threatened traditional doctrine. It was suppressed by the Vatican during his lifetime.

Despite its mixed reception, Teilhard's core intuition, that the emergence of consciousness represents the universe becoming aware of itself, and that this self-awareness is not an accident but the expression of a directional tendency in cosmic evolution, has been taken seriously by a number of scientists and philosophers. The biologist E.O. Wilson, the philosopher Thomas Nagel in his

2012 book Mind and Cosmos, and the complexity theorist Stuart Kauffman have all, in different ways and with different levels of commitment, explored the idea that the emergence of minds in the universe is not merely improbable but in some sense expectable or inevitable given the nature of the process.

Where the Evidence Points, Honestly

I said at the beginning of this chapter that I wanted to approach these questions with honesty. Here is my honest assessment of where the evidence points.

The evidence supports the claim that the universe has, over its history, produced structures of increasing organizational complexity, increasing capacity for information processing, and, eventually, increasing capacity for experience and self-reflection. This trajectory is real, documented, and quantifiable. Whether it constitutes a direction in the teleological sense, whether it implies that complexity and consciousness are goals toward which the universe is moving, is a different and harder question.

The honest answer is that we do not know. The evidence is consistent with Teilhard's vision of directed cosmic evolution, but it is also consistent with the view that the emergence of complexity and consciousness is a contingent product of particular initial conditions and local physical laws, without any deeper directional significance. The two interpretations are not distinguishable by currently available empirical means.

What I can say with more confidence is this: the metamorphic pattern, as it has been described in this book, consistently produces, when it succeeds, organizations of greater complexity and greater capability than the organizations that preceded them. The butterfly is more complex than the caterpillar. The post-supernova universe is chemically richer than the pre-supernova universe. The post-traumatic human, when growth occurs, has greater emotional complexity and relational capacity than the pre-traumatic one. The post-scientific-revolution civilization has greater capacity for understanding and manipulating the physical world than the pre-scientific-revolution one.

This is not a universal law. Metamorphic events often fail. The chrysalis sometimes dies. But when they succeed, the pattern is consistent: toward greater complexity, greater sensitivity, greater capacity for integration of experience. Whether this consistent tendency reflects a cosmic purpose, a physical law, or a happy accident, it is the pattern by which the universe has, over thirteen billion years, produced the beings who are now asking this question.

That fact alone, that metamorphosis has produced minds capable of wondering about metamorphosis, seems to me to be one of the most significant facts in the history of the universe. What we do with it is the subject of the book's final section.

XIV

Engineering Emergence

CRISPR, synthetic biology, and the power to design transformation

In January 2020, a team of researchers led by Josh Bongard at the University of Vermont and Michael Levin at Tufts University published a paper in the Proceedings of the National Academy of Sciences describing organisms they called xenobots: millimeter-scale living machines built from cells of the African clawed frog, Xenopus laevis, and designed by a computer algorithm to perform specific tasks.

The xenobots were not robots in the conventional sense. They were composed entirely of living cells, frog cells that under normal developmental circumstances would form skin or heart muscle, reprogrammed by the combination of algorithmic design and biological assembly to self-organize into forms that could move, carry loads, and, in a subsequent study published in 2021, spontaneously replicate by pushing loose cells in their environment into clusters that developed into new xenobots.

The xenobots are not genetically modified. The cells from which they are built have the normal Xenopus genome. What was changed was not the genome but the developmental context: the cells were removed from their normal developmental environment and placed in a new one that, combined with the algorithmic design of the structure, caused them to express their developmental programs in a radically different way, producing an organism with no analogue in evolutionary history.

Bongard and Levin described the xenobots as a novel living technology, and the description is apt. They represent something genuinely new in the history of biology: an organism whose form was designed by a computational algorithm, built from the cells of another organism, expressing developmental programs that evolved for entirely different purposes in an entirely different context. They are, in a precise sense, engineered emergence.

CRISPR and the Revolution in Genomic Editing

The development of CRISPR-Cas9 as a genomic editing tool is one of the most consequential scientific developments of the twenty-first century. The technology derives from a component of the bacterial immune system in which short repeated sequences in the bacterial genome, together with Cas proteins that can be guided to specific DNA sequences, allow bacteria to recognize and cut the DNA of invading viruses. In 2012, Jennifer Doudna of UC Berkeley and Emmanuelle Charpentier, then at the Helmholtz Centre for Infection Research, published the paper in Science that demonstrated how this system could be adapted as a programmable tool for editing DNA at specific, chosen locations.

The award of the 2020 Nobel Prize in Chemistry to Doudna and Charpentier recognized the transformative significance of this work. In the decade following that paper, CRISPR-based technologies moved into clinical and agricultural applications at a pace that outran the development of regulatory frameworks.

In December 2023, the United States Food and Drug Administration approved Casgevy, a CRISPR-based therapy for sickle cell disease developed by Vertex Pharmaceuticals and CRISPR Therapeutics. Casgevy works by editing the gene BCL11A in a patient's hematopoietic stem cells, reactivating the production of fetal hemoglobin, which does not carry the sickle cell mutation. In clinical trials, the treatment eliminated vaso-occlusive crises in 97 percent of treated patients. It does not manage the disease. It changes the molecular biology of the patient's blood cells in a way that makes the disease's primary mechanism inoperative. It is, in metabolic terms, a designed metamorphosis at the cellular level.

Synthetic Biology and Designed Organisms

Synthetic biology extends beyond editing existing genomes to designing and constructing novel biological systems. The global synthetic biology market was valued at approximately 13.9 billion US dollars in 2022 and projected to reach roughly 55.8 billion dollars by 2028, according to market analysis firm MarketsandMarkets. Applications span pharmaceutical biosynthesis, industrial chemistry, environmental remediation, and biological computing.

For the biology of metamorphosis specifically, synthetic biology opens the possibility of designing entirely novel metamorphic programs: developmental transitions between body plans that do not exist in any natural organism and could not have evolved through normal natural selection. More sophisticated versions, still speculative but not beyond foreseeable technological reach, would include therapeutic organisms designed to undergo a metamorphic transition within a patient's body, transforming from a delivery vehicle into a tissue-remodeling agent in response to specific disease signals.

The power to design emergence also carries a temptation that deserves naming: the temptation to skip the chrysalis. One of the most consistent findings about natural metamorphosis is that the dissolution phase cannot be bypassed without damaging the outcome. Researchers who interrupt the pupal phase of insect development invariably produce malformed adults. The dissolution does its necessary work, and that work takes the time it takes. When human beings design transformation processes, there is a consistent impulse to engineer rapid emergence while minimizing dissolution.

This impulse should be resisted on principled grounds, not merely pragmatic ones.

The dissolution phase is not a transitional inefficiency. It is the mechanism by which the old organization's resources and constraints are released and made available to the new organization. An emergence that has not been preceded by sufficient dissolution produces something neither fully the old form nor the new: functional in neither mode. The caterpillar removed from its chrysalis before the transformation is complete does not become a butterfly. It becomes something unable to be either.

The Three Ethical Axes

The questions raised by the power to design metamorphic programs can be organized around three axes.

The first is consent. For individual therapeutic applications like Casgevy, the patient consents. But engineered organisms released into environments, gene drives designed to propagate through wild populations, synthetic organisms designed to alter ecosystem chemistry: these cannot obtain the consent of the organisms and ecosystems they affect. The absence of consent mechanisms in ecological applications of engineered metamorphosis is among the most serious unresolved ethical problems in contemporary biotechnology.

The second is reversibility. Natural metamorphic processes at the organismal level are largely irreversible. Engineered metamorphic processes may or may not be reversible, and the question of whether designed transformation should be reversible, and if so how reversibility is to be built in and maintained, is one that current regulatory frameworks do not adequately address.

The third is what we might call the compression problem. The temptation to design transformation processes that minimize the dissolution phase and accelerate the emergence is not merely a technical error. It reflects a deep cultural discomfort with the chrysalis state, the period of apparent inactivity and vulnerability that precedes emergence. A biotechnology culture that cannot tolerate the chrysalis will consistently produce inadequate butterflies.

The lesson that the biology of metamorphosis offers to the practice of biotechnology is direct: respect the dissolution. This is not a counsel of passivity. It is a recognition that the most important work of transformation is often invisible, that it takes the time it takes, and that attempts to accelerate or bypass it typically produce outcomes that are less fully transformed than they would have been had the process been allowed to proceed at the pace the system required.

XV

Artificial Intelligence as Metamorphic Intelligence

The imaginal disc of civilization

In November 2020, researchers at DeepMind published a paper in Nature describing a system called AlphaFold 2. Protein folding, the problem of predicting the three-dimensional structure of a protein from its amino acid sequence alone, had been formally unsolved since it was clearly articulated in the late 1960s. The structure of a protein determines its function, and knowing the structure is essential for understanding how a protein works and for designing molecules that interact with it.

AlphaFold 2 solved it. In the 14th Critical Assessment of Protein Structure Prediction competition, the biennial benchmark for the field held in 2020, AlphaFold 2 produced predictions with median accuracy scores comparable to experimentally determined structures, far exceeding any previous computational approach. By July 2022, DeepMind had released structural predictions for more than two hundred million proteins, covering virtually every protein in the databases of sequenced genomes. The biochemist Venki Ramakrishnan, a Nobel laureate, described it as a complete solution to the protein folding problem.

The significance of AlphaFold 2 for this chapter's argument is not primarily the scientific achievement, remarkable as it is. It is what the achievement illustrates about the relationship between AI capabilities and the organizational structures of human knowledge. Structural biology had organized itself around the practical impossibility of rapid structural determination: expensive laboratory equipment, years of training, specialist facilities, proprietary databases. AlphaFold 2 did not merely make protein structure prediction faster. It changed the fundamental assumptions around which an entire field had organized itself.

This is what imaginal discs do.

The Imaginal Disc Hypothesis

In biological metamorphosis, imaginal cells are the elements within the larval body that carry the blueprint for the adult form. Present from the beginning but dormant, they do not express their identity or function while the larval organization persists. The larval organization does not accommodate them because the organizational form they represent is incompatible with the larval form. Their activation requires, and also accelerates, the dissolution of the larval organization.

The hypothesis I want to propose, with appropriate tentativeness, is that artificial intelligence functions in the current metamorphic transition of human civilization as a large-scale imaginal disc: a set of capabilities and organizational possibilities that are present within the current civilization but that the current civilization's organizational structures cannot fully accommodate. The emergence of these capabilities is creating pressure on existing institutions, economic structures, knowledge systems, and social arrangements in ways that are characteristic of a chrysalis event.

This is not a prediction about what AI will do or become. It is an observation about what it is doing, now, to the civilization in which it is being developed. And what it is doing is dissolving organizational forms that have been stable for generations.

What the Dissolution Looks Like

The World Economic Forum's Future of Jobs Report 2023 estimated that 69 million new jobs would be created globally over the 2023 to 2027 period while 83 million would be displaced, a net displacement of 14 million jobs in five years. Behind this net figure lies a more disruptive reality: the displaced jobs and the created jobs are not in the same industries, do not require the same skills, and are not geographically co-located. The transition between them is a liminal phase that the people living through it experience as personal dissolution, not civilizational reorganization.

Beyond labor markets, the dissolution is visible in the institutions of professional knowledge. Large language models and specialized AI systems are demonstrating capabilities in legal reasoning, medical diagnosis, scientific literature synthesis, and financial analysis that challenge the guild-like structures through which these professions have controlled access to expertise. This does not mean the professions will disappear. It means the organizational structures through which they have been practiced will need to metamorphose. The value they create, expert judgment applied to complex human situations, is not disappearing. The structures that have historically housed that value are being dissolved.

The educational system is experiencing the same pressure. An educational model designed to transmit information and train procedural skills is being rendered partially obsolete by systems that can access, synthesize, and apply information at a scale and speed that no individual can match. What education must become, in the emerging order, is something different from what it has primarily been: not primarily information transmission but the cultivation of judgment, creativity, ethical reasoning, and the relational capacities that remain distinctively human. This requires a metamorphic transition in educational philosophy and practice that most educational institutions have not yet seriously begun.

The Civilizational Immune Response

The reaction of existing institutions to the emergence of AI capabilities has many characteristics of an immune response attacking imaginal cells. Professional associations have moved to restrict AI use in their domains. Educational institutions have focused energy on preventing AI-assisted academic work. Media organizations have pursued legal strategies to limit AI training on their archives. These responses are understandable expressions of the interest of institutions in maintaining their current organizational form. They are also, in metamorphic terms, likely to be temporary.

The point is not that the concerns motivating these responses are illegitimate. Copyright, quality standards, and professional accountability are real values. The point is that, as in biological metamorphosis, the immune response cannot ultimately suppress the imaginal cells once they have reached sufficient critical mass. The question is not whether the transition will occur but whether it can be navigated in a way that preserves the genuine values that the dissolving structures embodied while releasing the constraints that made those structures incompatible with the emerging organization.

The civilizational transitions that have succeeded, as discussed in Chapter Ten, have been characterized by the survival and incorporation of the genuine values of the dissolving structures into the new synthesis. The genuine value of professional medical expertise is not the guild structure that licenses it. It is the commitment to patient welfare and evidence-based clinical judgment. An AI-augmented medical system that preserves these values while releasing the guild structure's constraints on access and cost would represent genuine metamorphic success.

What AI Is Forcing Us to Discover

The most important thing that artificial intelligence is doing to human civilization is not the economic disruption, significant as that is. It is the philosophical pressure it is applying to the question of what human beings actually are and what, specifically, they are for.

For several centuries, the dominant answer to this question in Western culture has been implicitly cognitive: human beings are rational animals, and their distinctive value lies in their capacity for reasoning, information processing, and the application of intelligence to problems. This answer was not wrong, but it was incomplete, and it became culturally dominant in part because the cognitive capacities it emphasized were scarce and therefore economically valuable.

Artificial intelligence is making certain cognitive capacities abundant. When pattern recognition, information retrieval, and procedural reasoning can be performed by AI systems at a fraction of the human cost, the human beings who have defined their value primarily through these capacities face a challenge that is not merely economic but existential: what are we for, if not primarily these things?

The metamorphic framing suggests that this question, experienced as a threat, is actually a revelation. Artificial intelligence is not replacing human intelligence. It is forcing human intelligence to discover what it actually is, beyond the cognitive functions that machines can now replicate. What it is, at least in part, is relational, embodied, morally situated, and experiential in ways that current AI systems are not. The cultivation of these dimensions of human capacity is the work of the chrysalis that AI is precipitating. It is not comfortable work. But it is, if the metamorphic pattern holds, the precondition for something genuinely new.

Part Five: The Future of Metamorphosis

XVI

Can Humanity Collectively Metamorphose?

The greatest question of the twenty-first century

There is a photograph taken by the Apollo 17 crew on December 7, 1972, at a distance of approximately 45,000 kilometers from Earth. It shows the full disk of the planet against the black of space: Africa and Antarctica clearly visible, the Arabian Peninsula, a partial view of Asia, swirling cloud patterns over the Southern Ocean. The photograph is known as The Blue Marble, and it became one of the most widely reproduced photographs in history.

What the photograph did, for the first time in human history, was make visible something that had always been true but had never been directly perceivable: that the Earth is a finite, bounded, isolated system. Not infinite. Not self-renewing beyond any possible demand. Not separate from its human inhabitants in the way that a stage is separate from the actors who perform on it. A single integrated system of which the human civilization developing on its surface is both a product and a major determinant.

The implications of this visibility have not been adequately metabolized in the fifty-plus years since the photograph was taken. They are the context in which the question of this chapter must be asked. Can humanity collectively metamorphose? And the answer, if one is possible at all, must be given against the background of a planetary system that is being altered by human activity at rates and scales that have no precedent in the geological record outside of mass extinction events.

This is not, as it may appear, an introduction to a chapter about climate change. It is an introduction to a chapter about the conditions that determine whether a metamorphic event produces genuine transformation or collapse. The planetary context is one of those conditions.

The Evidence That We Are Mid-Dissolve

Several converging lines of evidence suggest that human civilization is currently in the dissolution phase of a metamorphic event of civilizational scale.

The first line of evidence is institutional. The major institutions through which modern civilization has been organized, the nation-state system established at Westphalia in 1648, the international economic order established at Bretton Woods in 1944, the multilateral security architecture established in 1945, are under increasing strain from challenges that they were not designed to address. Climate change, global pandemics, artificial intelligence, mass migration, and transnational digital communication all represent problems whose scale and nature exceed the capacity of the existing institutional architecture to manage.

The second line is epistemic. The information ecosystem through which modern democratic societies have organized their shared knowledge and collective decision-making, the combination of professional journalism, peer-reviewed science, and representative political deliberation, is fragmenting under the pressure of digital communication and the economic disruption of advertising-funded media. The result is not merely political polarization, real as that is, but something more fundamental: the erosion of shared epistemic frameworks within which disagreement can be meaningfully conducted.

The third is ecological. The Holocene, the roughly twelve-thousand-year period of unusual climatic stability within which all human civilization developed, is ending. The concentration of atmospheric carbon dioxide in

March 2024 stood at 425.22 parts per million, according to measurements from the Mauna Loa Observatory in Hawaii, the highest level in at least three million years. The consequences of this and related changes to the Earth system are already underway and will intensify over the coming decades regardless of future emissions trajectories.

These are not the only lines of evidence that a dissolution is underway, but they are sufficient to establish that the organizational forms through which human civilization has operated for the past several centuries are under fundamental stress. The question is not whether a metamorphic transition is occurring. The question is what kind.

The Five Conditions Revisited

In Chapter Ten, analyzing civilizational metamorphosis historically, five conditions were proposed as determinants of whether civilizational dissolution produces genuine metamorphosis or collapse. It is worth applying these conditions to the current situation.

The first condition was the survival of imaginal cells: institutions, practices, and knowledge systems that carry the blueprint for a new organization. The current civilization is generating imaginal cells in significant numbers. The rapid development of renewable energy technologies represents an imaginal disc carrying the blueprint for an energy system that does not depend on the combustion of geological carbon. In 2023, the International Energy Agency reported that solar power added more capacity globally than all other electricity sources combined for the second consecutive year. Distributed renewable generation, long-horizon ecological management, AI-augmented scientific research, and new models of democratic deliberation are all candidates for civilizational imaginal discs. The imaginal cells exist. The question is whether they will survive and connect.

The second condition was sufficient dissolution: the existing organizational structures must be sufficiently disrupted to release the resources and organizational space that the new synthesis requires. The current situation appears to meet this condition, perhaps more fully than is comfortable. The dissolution of institutional authority, epistemic frameworks, and economic models is, in many respects, already well advanced. The risk is not insufficient dissolution but dissolution that is too rapid and too broad to allow the imaginal cells to organize into a coherent new structure before the conditions necessary to support them are destroyed.

The third condition was connection among imaginal elements. This is where the current situation is most uncertain. The imaginal cells of a new civilization exist, but their connection to each other, and their ability to generate the kind of coordinating chemical environment that in biological metamorphosis allows imaginal cells to overcome the immune response and begin organizing the adult form, is limited. The organizational infrastructure for connecting innovative solutions across sectors, geographies, and knowledge domains does not yet exist at the scale required.

The fourth condition was time. This is the most troubling variable in the current assessment. Biological metamorphosis takes the time it takes, and cannot be safely accelerated. Civilizational metamorphosis has historically taken centuries. The ecological and geophysical pressures currently operating on human civilization are not compatible with a centuries-long timeline for organizational transition. The carbon budget remaining for limiting global warming to 1.5 degrees Celsius above pre-industrial levels, according to the Intergovernmental Panel on Climate Change's 2021 assessment, was approximately 300 to 500 gigatons of carbon dioxide equivalent as of 2020, at current emission rates a window of roughly seven to twelve years. The metamorphic transition must occur on a timeline that may be shorter than civilizational metamorphoses have historically required.

The fifth condition was collective willingness to release the previous organization's most valuable features while preserving its most essential ones. This condition is currently being contested. There are powerful forces within the current civilization that are committed to the preservation of the existing organizational forms, particularly the fossil fuel economy, the nation-state system of sovereignty, and the consumerist model of economic growth. There are also significant forces oriented toward transition. Which prevails, and on what timeline, is genuinely uncertain.

The Role of the Individual Imaginal Cell

In biological metamorphosis, individual imaginal cells do not know that they are building a butterfly. They respond to their local biochemical environment, expressing their developmental programs in response to the signals they receive. The butterfly is the emergent product of billions of individual cell behaviors, none of which has access to the overall plan.

The same is true of civilizational metamorphosis. No individual, and no institution, has access to the overall plan of what is being built. Each acts in response to its local context, its values, its capacities, and the signals it receives from the surrounding environment. The civilization that emerges from the current transformation will be the product of billions of individual choices and billions of institutional decisions, none of which is fully aware of the others.

This is not a counsel of helplessness. It is a description of how large-scale transformation actually works, and it implies something important about what any individual person can do.

The imaginal cells that succeed in biological metamorphosis are those that express their identity clearly, connect with other imaginal cells, and resist the immune response long enough to reach critical mass. For an individual human being in a civilization undergoing metamorphic transition, the analogous behaviors are: to identify and develop the capacities that the new organization will need rather than the ones that the old organization has valued; to connect deliberately with others who are doing the same; and to persist in that expression in the face of the resistance that the dissolving culture will mount against it.

This is neither optimistic nor pessimistic. It is metamorphically realistic. The chrysalis may fail. Many do. The conditions facing the current civilization are genuinely severe, and the outcome is genuinely uncertain. But the uncertainty is not fixed. It is responsive to what happens inside the chrysalis, including what individual imaginal cells do.

The Personal and the Planetary

This book began with a moth emerging from a cocoon, watched by a person who was themselves in the middle of a chrysalis event. It seems appropriate, at the end, to return to that pairing.

The connection between personal transformation and civilizational transformation is not merely metaphorical. It is structural. The same five conditions that determine whether a civilization emerges from dissolution, survival of imaginal cells, sufficient dissolution, connection among imaginal elements, appropriate time, and collective willingness to release the old form, operate at the personal scale. The same chrysalis paradox, the invisibility of the most active transformation, applies to both. The same imaginal disc logic, the activation of capacities that were present but dormant, describes both.

This means that the work of personal metamorphosis, genuinely undertaken, is not separate from the work of civilizational transition. A person who has learned to navigate the chrysalis phase of their own life with skill and honesty, who has developed the capacity to recognize dissolution without collapsing into it, to trust the process of emergence without forcing it, to acknowledge what must be released without denying the loss: such a person is better equipped to function as an imaginal cell in a civilization undergoing the same transition.

The cultivation of metamorphic literacy, at the personal scale, is among the most practically significant things a person can do in a civilizational chrysalis. Not because it solves the large-scale problems, it does not. But because the large-scale problems will not be solved by people who cannot tolerate their own dissolution. The imagination required to build genuinely new organizational forms is available only to people who have learned, in some measure, to release their attachment to the organizational forms that preceded them.

The moth I watched emerging that autumn evening flew into the dark before I could see where it was going. I have thought about that since. The moth did not need to know where it was going. It needed only to have completed the transformation that made flight possible.

We are building wings. It is difficult to see them clearly from inside the chrysalis. But the cells that will carry us are already active, already organizing, already learning to resist the suppression that the dissolving order will continue to mount against them. They have been doing so, in one form or another, for as long as living systems have been responding to conditions that exceed their current capacity for incremental change.

That is a very long time. And it has produced, among other things, the beings who are now reading this sentence and wondering what they might become.

Conclusion

The Art of Metamorphic Living

There is a practice among Zen Buddhist teachers of posing koan to students: riddles or questions designed not to be answered by logical analysis but to precipitate a shift in how the student understands the nature of mind and reality. The most famous is: What is the sound of one hand clapping? Less famous but more relevant to the subject of this book is one attributed to the thirteenth-century teacher Dogen Zenji: Before enlightenment, chop wood, carry water. After enlightenment, chop wood, carry water.

The question implicit in Dogen's statement is: if the activity is the same before and after, what has changed? The answer, which cannot be given in a single sentence but which the preceding pages have been attempting to approach from multiple directions, is: everything about how the activity is inhabited. Not the wood. Not the water. Not the chopping or the carrying. The quality of presence that meets them.

Metamorphic literacy does not give you a different life. It gives you a different relationship to the life you have, including the parts of it that are currently dissolving.

This conclusion draws together the book's argument into seven principles for metamorphic living. They are not prescriptions. They are observations drawn from the evidence of biology, cosmology, psychology, and history, tested against the grain of lived experience, and offered with the recognition that knowing them and practicing them are entirely different things.

The Seven Principles

One: Recognize the chrysalis when you are in it.

The chrysalis paradox means that the experience of being in a metamorphic event resembles the experience of failure or deterioration. The structures that previously organized your life are dissolving. The clarity that characterized your previous phase is gone. You feel less capable, less certain, less yourself than you were before.

The first skill of metamorphic living is learning to distinguish dissolution from deterioration. Deterioration is characterized by a reduction in the quality and quantity of engagement with life. The chrysalis is characterized by a radical reorganization of what engagement means. Things that mattered before no longer matter in the same way. Things that seemed peripheral are becoming central. The old organization is losing coherence while the new one has not yet assembled itself.

Recognizing this does not make it less uncomfortable. But it makes it legible. And legibility is the beginning of navigation.

Two: Do not force emergence.

The most common mistake made by people in a chrysalis phase, and by those who care about them, is the attempt to force or accelerate emergence: to push for resolution before the dissolution has done its work. The butterfly whose chrysalis is opened prematurely cannot complete the pumping of hemolymph into its wings. The wings do not expand. Forced emergence is not no emergence. It is partial emergence, and partial emergence is, in many respects, worse than waiting.

The dissolution phase takes the time it takes. The question to ask is not how do I make this end sooner, but how do I inhabit this phase with enough integrity and patience that what emerges from it is genuinely new rather than simply a hurried reconstruction of what was there before.

Three: Trust the imaginal cells.

In every chrysalis event, there are elements of the new organization that are already present: nascent values, emerging capacities, relationships and practices that are beginning to feel more alive than the ones that are dissolving. These are the imaginal cells of your transformation. They are often small and they often seem peripheral or even threatening to the part of you that is still identified with the old organization.

The practice of trusting the imaginal cells means attending to what is becoming alive in you even when it is small and unclear. It means investing in the emerging capacities rather than directing all resources toward maintaining the dissolving ones. When the dissolving organization mounts its immune response and registers the new elements as foreign, persist in the development of those elements rather than suppressing them in the service of a coherence that is already failing.

Four: Grieve what is dissolving with full honesty.

The temptation in transformation is to minimize the reality of what is being lost, to comfort oneself with the certainty that what is coming is better than what is going, and to rush past the grief in the direction of the emergence. This is a mistake.

What is dissolving in a metamorphic event is real, and it deserves honest grief. The person you were before the transformation was genuinely you. The relationships organized around that person were genuine relationships. Metamorphic living includes the practice of grieving fully and honestly, not as a stage to be moved through and left behind, but as an ongoing relationship with the selves that have preceded you.

Five: Maintain accountability through transformation.

You are the same person, in all the ways that matter ethically, as the person who made the commitments that preceded your transformation. The fact that you have changed does not dissolve those commitments. It changes the context in which they must be honored, and it may, in some cases, make honoring them in their original form impossible. But it does not eliminate the obligation to acknowledge them, to address the people affected by your transformation with honesty and respect, and to make whatever repair is possible.

Transformation without accountability is not metamorphosis. It is flight with better language.

Six: Develop tolerance for the in-between.

The liminal phase of transition is the most generative and the most dangerous. It is the moment at which old certainties are gone and new ones have not yet formed, and it is also the moment at which the creative possibilities of transformation are most fully open. The person who can tolerate this in-between, who can live in the uncertainty of the chrysalis without collapsing the process by forcing premature resolution, has access to a quality of creative possibility that the stable phases of life do not offer.

Contemplative practices, good therapeutic relationships, rigorous creative work, and communities of honest inquiry can all support this development. None of them eliminates the difficulty. All of them can increase the capacity to stay with it.

Seven: Connect your transformation to something larger.

The imaginal cells of a biological metamorphosis succeed in building the adult organism because they connect with other imaginal cells. The isolated imaginal cell cannot build a butterfly. The connected community of imaginal cells can. The same principle operates at the human scale.

Personal transformation is most fully meaningful when it is connected to a transformation that exceeds the personal. The person who understands their own chrysalis event as part of a larger metamorphic process inhabits their transformation differently. It has stakes that exceed the personal. It is not merely self-improvement. It is imaginal cell activation.

The universe has been practicing metamorphosis for thirteen point eight billion years. It has produced, in that time, stars and galaxies and planets and life and minds. It has done all of this through the repeated application of the same structural logic: dissolution, reorganization, emergence. You are a product of that process and a participant in its continuation.

The chrysalis you are in, whoever you are and whatever form it takes, is not separate from that process. It is an expression of it.

Afterword

The moth I described in the Preface emerged from its cocoon on a Wednesday evening in early October, in a field on the edge of a nature reserve in southern England. It was a member of Lasiocampa quercus, the oak eggar moth, a species common enough in hedgerows and woodland edges across much of Europe. The cocoon was attached to a dried stem of Yorkshire fog grass. The emergence took fourteen minutes from the first visible split in the cocoon wall to the point at which the moth pumped its wings and became still.

I have not returned to that field, though it is not far from where I now live. I have not needed to. The moth has been with me every day since, not as a memory but as a question that the preceding pages have been my best attempt to answer.

I am not the same person who stood in that field. That person did not know the things that are in this book, and did not have the relationships, the clarity, or the capacity for certain kinds of work that I have now. In the years since that evening, I have also lost things that were real and that I grieve. A form of innocence. A certain quality of easy certainty. Several relationships that belonged to the person I was before the transformation and that did not survive it.

I write none of this to claim that my transformation was successful or complete. Metamorphic events do not conclude the way books do. They leave the organism in a new phase from which further transformation will, in time, proceed. I am not a butterfly who has finished metamorphosing. I am a butterfly who is, for now, flying, and who knows that the conditions of that flight will change, and that what comes next will require its own dissolution.

What the writing of this book has given me, beyond the specific knowledge it contains, is something harder to name: a relationship to my own changing that is less frightened than it was. Not because transformation has become comfortable, but because it has become legible.

You are reading this, whoever you are, because something brought you to a book with this title. I do not know what that something is. I hope the book has served it.

The field is still there. The oak eggar moths are still emerging, in October, from their cocoons attached to the dried grass stems. The universe is still converting hydrogen to helium in a hundred billion billion stars, still distributing heavy elements through supernovae, still building complexity from the wreckage of previous organizations.

It will be doing this long after both of us are gone, having dissolved our molecules back into the pool from which the next generation of imaginal cells will be built.

That is not a terrifying thought. It is the most honest account I know of how everything that matters came to exist.

Acknowledgments

A book that crosses as many disciplinary boundaries as this one depends on the generosity of specialists willing to help a non-specialist understand their field with sufficient depth and accuracy to write about it responsibly.

For the biological material in Part One, I am indebted to the published work of Lynn Riddiford of the University of Washington, whose four decades of research on the hormonal regulation of insect metamorphosis represent the finest sustained investigation of the subject in the scientific literature. The work of Douglas Blackiston and Martha Weiss, whose 2008 study of memory transfer through metamorphosis in Manduca sexta is one of the most elegant experimental designs in recent behavioral biology, was essential to Chapter One. Michael Levin's research program at Tufts University, encompassing both the xenobot work and the planarian experiments, shaped Chapters Two and Fourteen. The research of Toshiyuki Nakagaki and colleagues on Physarum polycephalum opened Chapter Two.

For the cosmological and physical material in Part Two, I have relied on the published work of Lawrence Krauss, whose clarity in communicating astrophysics without losing its actual content has been a model for this book's ambitions. The Planck satellite mission data, released in 2018 by the European Space Agency, provided the figures for dark matter and dark energy composition cited in Chapter Six. Vera Rubin's original papers on galactic rotation curves informed the historical narrative of that chapter.

For the neuroscientific and psychological material in Part Three, I have drawn on Eleanor Maguire's taxi driver studies, Fred Gage's work on adult hippocampal neurogenesis, and Sara Lazar's cortical thickness research on meditators. Jill Bolte Taylor's memoir My

Stroke of Insight was indispensable for Chapter Seven. Richard Tedeschi and Lawrence Calhoun's framework for post-traumatic growth informed the analysis in Chapter Eight. George Bonanno's longitudinal research on resilience and grief provided the empirical foundation for the critique of the stage model. David Kessler's work on the sixth stage of grief, published in Finding Meaning in 2019, extended that chapter's analysis.

The anthropological material in Chapter Eight draws on Arnold van Gennep's Les Rites de Passage of 1909 and Victor Turner's subsequent development of the liminality concept. Jennifer Freyd's betrayal trauma theory informed the ethical analysis in Chapter Twelve.

The woman called Diane in Chapter One spoke with me about her experience of illness and recovery in an interview conducted in 2019. She approved the passages about her before publication. Elena, whose story opens Chapter Twelve, is a composite drawn from multiple people whose transformations I have known over many years.

To every person who has, in conversation or simply by living their transformation with honesty and courage in my presence, contributed to the thinking in these pages: this book is, in part, yours.

Glossary of Key Terms

Anatta
The Buddhist concept of non-self, denying the existence of a permanent, unchanging self that persists through time and experience. What we call the self is a flowing process of physical and mental events conventionally designated by a name and a body.

Aufhebung
A German philosophical term used by Hegel, meaning simultaneously cancellation, preservation, and elevation. Describes how a previous stage of development is cancelled in its immediate form while being preserved and elevated in the new form that supersedes it.

Chrysalis Paradox
The phenomenon in which maximal internal transformation coincides with minimal external expression. The chrysalis, during the most active phase of metamorphic reorganization, appears from outside to be inert or dead.

Chrysalis Phase
The period within any metamorphic event during which the previous organization has been substantially dissolved and the new organization has not yet fully assembled. Characterized by ambiguity, vulnerability, and intense internal reorganization.

Dedifferentiation
The process by which a fully differentiated cell reverts to a less differentiated state before proliferating and redifferentiating to produce a new cell type. Central to the regenerative biology of organisms like the axolotl.

Default Mode Network

Brain regions that show coordinated activity during rest, associated with self-referential thought and autobiographical narrative. Shows altered connectivity patterns in long-term meditators.

Ecdysteroid
A class of steroid hormones that drives the molting process in arthropods. In coordination with juvenile hormone, ecdysteroid signaling determines the identity of each successive molt in holometabolous insects.

Emergence
The phenomenon in which properties appear at one level of organization that were not predictable from the components at a lower level. Water is wet; individual water molecules are not. Consciousness arises from neural activity; individual neurons do not appear to be conscious.

Entelechy
An Aristotelian concept describing the principle by which a thing contains within itself the form of what it is striving to become. The acorn has the oak as its entelechy: an internal organizing principle directing development toward a specific end state.

Epigenetics
The study of heritable changes in gene expression that do not involve changes in the DNA sequence itself. Epigenetic modifications can alter gene expression patterns and in some cases be transmitted across generations.

Histolysis
The systematic enzymatic breakdown of larval tissues during the pupal phase of holometabolous insect development. The caterpillar secretes enzymes that dissolve its own muscles, fat body, and digestive tract into an undifferentiated biological fluid.

Holometabolism

The developmental strategy of complete metamorphosis, involving four distinct life stages: egg, larva, pupa, and adult. The transition from larval to adult form occurs during the pupal stage. Approximately 85 percent of all insect species are holometabolous.

Hox Genes
A family of transcription factors whose DNA-binding domains are highly conserved across animal species, performing equivalent developmental functions in organisms as different as flies and vertebrates. Master regulatory sequences that specify body segment identity.

Imaginal Cell
A cell or group of cells in a holometabolous insect larva that remains dormant during the larval period but carries the genetic programming to produce adult structures during metamorphosis. Used throughout this book as a conceptual metaphor for any element within a system that carries the blueprint for a future organizational form.

Imaginal Disc
A cluster of imaginal cells, typically numbering ten thousand to fifty thousand cells, each committed to producing a specific adult structure such as a wing, leg, or eye.

Juvenile Hormone
A hormone produced in insects that, when present during an ecdysteroid-driven molt, causes the resulting molt to be larval rather than pupal or adult. The absence of juvenile hormone during an ecdysteroid peak initiates metamorphosis.

Liminal Phase
From the Latin limen, meaning threshold. In anthropology, the middle phase of a rite of passage, in which the individual is no longer what they were and not yet what they will become. Developed by Arnold van Gennep and extended by Victor Turner.

Logos

In Heraclitean philosophy, the rational principle of transformation, the law by which opposites give way to each other and all things flow. The organizing process that persists through and directs transformation.

Metamorphic Literacy
The capacity to recognize a metamorphic event when one is inside it, to distinguish dissolution from deterioration, and to respond appropriately to each phase of the transformation. A central practical aim of this book.

Neoteny
The retention of juvenile characteristics in the adult form of an organism. The axolotl is neotenic, retaining larval features throughout its sexually mature adult life.

Neuroplasticity
The capacity of the brain to change its structure and function in response to experience, learning, injury, or disease. Established as a mainstream scientific concept primarily through the work of Michael Merzenich.

Phase Transition
A qualitative reorganization of a system's structure triggered by a quantitative change in a control parameter. The transition of water from liquid to ice is a first-order phase transition.

Post-Traumatic Growth
Positive psychological changes reported by some individuals following severely challenging life circumstances, including increased personal strength and relational depth. Coined by Richard Tedeschi and Lawrence Calhoun in 1996.

Process Philosophy
A philosophical tradition associated with Alfred North Whitehead that takes process, event, and becoming as the fundamental ontological categories rather than substance and being.

Superposition

In quantum mechanics, the state of a quantum system in which it exists simultaneously in multiple possible states until a measurement or interaction collapses it into a single definite state.

Xenobots
Millimeter-scale living machines built from cells of the African clawed frog and designed by a computer algorithm. Described in a 2020 paper by Josh Bongard and Michael Levin. The first designed organisms whose form was algorithmically generated rather than evolutionarily selected.

Notes and References

The following notes provide citations for specific claims in each chapter along with occasional commentary on contested or complex points.

Introduction

The description of SN 1054 draws on Stephenson, F.R. and Green, D.A. Historical Supernovae and Their Remnants. Oxford University Press, 2002. The Crab Pulsar rotation rate is from the Australia Telescope National Facility Pulsar Catalogue, publicly available at atnf.csiro.au.

Manduca sexta metamorphosis: Riddiford, L.M. How does juvenile hormone control insect metamorphosis and reproduction? General and Comparative Endocrinology, 179(3), 477-484, 2012.

The interview with Diane was conducted in London in October 2019. She reviewed and approved all passages in which she appears.

Chapter 1: The Creature That Dissolves Itself

The imaginal disc cell count and Monarch butterfly wing data: Reppert, S.M., Gegear, R.J., and Merlin, C. Navigational mechanisms of migrating monarch butterflies. Trends in Neurosciences, 33(9), 399-406, 2010.

Memory transfer through metamorphosis: Blackiston, D.J., Silva Casey, E., and Weiss, M.R. Retention of memory through metamorphosis: Can a moth remember what it learned as a caterpillar? PLOS ONE, 3(3), e1736, 2008. The 77 percent retention figure is from this paper.

Imaginal cell immune response: Discussed in Carroll, S.B. Endless Forms Most Beautiful. W.W. Norton, 2005, drawing on foundational immunological and developmental biology literature.

Chapter 2: Beyond the Butterfly

Slime mold maze: Nakagaki, T., Yamada, H., and Toth, A. Maze-solving by an amoeboid organism. Nature, 407(6803), 470, 2000.

Tokyo rail network: Tero, A., et al. Rules for biologically inspired adaptive network design. Science, 327(5964), 439-442, 2010.

Planarian memory through head regeneration: Shomrat, T. and Levin, M. An automated training paradigm reveals long-term memory in planarians and its persistence through head regeneration. Journal of Experimental Biology, 216(20), 3799-3810, 2013.

Bacillus subtilis sporulation: Errington, J. Bacillus subtilis sporulation: Regulation of gene expression and control of morphogenesis. Microbiological Reviews, 57(1), 1-33, 1993.

Chapter 3: The Metamorphic Genome

Nobel Prize 1995: Announced October 9, 1995, Royal Swedish Academy of Sciences, for Nüsslein-Volhard, Wieschaus, and Lewis.

CRISPR-Cas9: Jinek, M., et al. A programmable dual-RNA-guided DNA endonuclease in adaptive bacterial immunity. Science, 337(6096), 816-821, 2012.

FDA approval of Casgevy: US Food and Drug Administration Press Release, December 8, 2023.

Chapter 4: Stars Are Creatures Too

Two trillion galaxies: Conselice, C.J., et al. The evolution of galaxy number density at $z<8$ and its implications. The Astrophysical Journal, 830(2), 83, 2016.

Stellar nucleosynthesis: Burbidge, E.M., Burbidge, G.R., Fowler, W.A., and Hoyle, F. Synthesis of the elements in stars. Reviews of Modern Physics, 29(4), 547, 1957.

The Krauss paraphrase draws on Krauss, L.M. A Universe from Nothing. Free Press, 2012.

Chapter 5: Quantum Metamorphosis

Young's double-slit experiment with electrons: Jönsson, C. Electron diffraction at multiple slits. American Journal of Physics, 42(1), 4-11, 1974.

Bell inequality violations: Aspect, A., Grangier, P., and Roger, G. Experimental realization of Einstein-Podolsky-Rosen-Bohm gedanken experiment. Physical Review Letters, 49(2), 91, 1982.

Whitehead process philosophy: Whitehead, A.N. Process and Reality. Macmillan, 1929.

Chapter 6: The Dark Chrysalis

Planck 2018 data: Planck Collaboration. Planck 2018 results VI: Cosmological parameters. Astronomy and Astrophysics, 641, A6, 2020.

Vera Rubin rotation curves: Rubin, V.C., Ford, W.K., and Thonnard, N. Extended rotation curves of high-luminosity spiral galaxies. The Astrophysical Journal, 225, L107-L111, 1978.

Hard problem of consciousness: Chalmers, D.J. Facing up to the problem of consciousness. Journal of Consciousness Studies, 2(3), 200-219, 1995.

Chapter 7: The Brain That Remakes Itself

Taylor, J.B. My Stroke of Insight. Viking, 2008.

London taxi driver study: Maguire, E.A., et al. Navigation-related structural change in the hippocampi of taxi drivers. PNAS, 97(8), 4398-4403, 2000.

Adult neurogenesis: Eriksson, P.S., et al. Neurogenesis in the adult human hippocampus. Nature Medicine, 4(11), 1313-1317, 1998.

Post-traumatic growth prevalence review: Shakespeare-Finch, J. and Lurie-Beck, J. A meta-analytic clarification of the relationship between posttraumatic growth and symptoms of PTSD. Journal of Anxiety Disorders, 28(2), 223-229, 2014.

Meditation cortical thickness: Lazar, S.W., et al. Meditation experience is associated with increased cortical thickness. NeuroReport, 16(17), 1893-1897, 2005.

Chapter 8: Grief, Loss, and the Cocoon You Did Not Choose

Cardiac risk after bereavement: Mostofsky, E., et al. Risk of acute myocardial infarction after the death of a significant person. Circulation, 125(3), 491-496, 2012.

Bonanno resilience: Bonanno, G.A. The Other Side of Sadness. Basic Books, 2009.

Prolonged grief disorder prevalence: Lundorff, M., et al. Prevalence of prolonged grief disorder in adult bereavement. Journal of Affective Disorders, 212, 138-149, 2017.

Kessler sixth stage: Kessler, D. Finding Meaning: The Sixth Stage of Grief. Scribner, 2019.

Van Gennep: Van Gennep, A. The Rites of Passage. University of Chicago Press, 1960 [1909]. Turner, V. The Ritual Process. Aldine, 1969.

Chapter 9: Who Are You After?

Parfit teleportation: Parfit, D. Reasons and Persons. Oxford University Press, 1984.

McAdams narrative identity: McAdams, D.P. The psychology of life stories. Review of General Psychology, 5(2), 100-122, 2001.

Winnicott and the object: Winnicott, D.W. The use of an object and relating through identifications. In Playing and Reality. Tavistock, 1971.

Chapter 10: Civilizations in the Chrysalis

Fall of Rome: Ward-Perkins, B. The Fall of Rome and the End of Civilization. Oxford University Press, 2005.

Corpus Juris Civilis: Commissioned 529-534 CE under Emperor Justinian I.

Copernicus: De Revolutionibus Orbium Coelestium. Nuremberg: Petreius, 1543.

Chapter 11: Is Change Real?

Heraclitus Fragment 91: Kirk, G.S. Heraclitus: The Cosmic Fragments. Cambridge University Press, 1954.

Hegel Aufhebung: Hegel, G.W.F. Phenomenology of Spirit. Translated by A.V. Miller. Oxford University Press, 1977 [1807].

Whitehead: Process and Reality. Macmillan, 1929.

Chapter 12: The Ethics of Becoming

Nietzsche: Thus Spoke Zarathustra. In The Portable Nietzsche, translated by Walter Kaufmann. Viking Press, 1954 [1883-1885].

Alcoholics Anonymous: Alcoholics Anonymous. 4th ed. A.A. World Services, 2001 [1939].

Betrayal trauma: Freyd, J.J. Betrayal trauma: Traumatic amnesia as an adaptive response to childhood abuse. Ethics and Behavior, 4(4), 307-329, 1994.

Chapter 13: Does Metamorphosis Have a Direction?

Chaisson energy rate density: Chaisson, E.J. Cosmic Evolution. Harvard University Press, 2001.

Teilhard: The Phenomenon of Man. Translated by Bernard Wall. Harper, 1959 [1955].

Big Bang nucleosynthesis: Alpher, R.A., Bethe, H., and Gamow, G. Physical Review, 73(7), 803-804, 1948.

Chapter 14: Engineering Emergence

Xenobots: Kriegman, S., et al. A scalable pipeline for designing reconfigurable organisms. PNAS, 117(4), 1853-1859, 2020.

Casgevy FDA approval: FDA Press Release, December 8, 2023.

Synthetic biology market: MarketsandMarkets Research Report, 2022.

Chapter 15: Artificial Intelligence as Metamorphic Intelligence

AlphaFold 2: Jumper, J., et al. Highly accurate protein structure prediction with AlphaFold. Nature, 596(7873), 583-589, 2021.

WEF Future of Jobs 2023: World Economic Forum. The Future of Jobs Report 2023.

IEA solar capacity: International Energy Agency. World Energy Outlook 2023.

Chapter 16: Can Humanity Collectively Metamorphose?

Apollo 17 photograph: NASA Goddard Space Flight Center. AS17-148-22727, December 7, 1972.

CO2 concentration 425.22 ppm: Scripps Institution of Oceanography, Mauna Loa Observatory dataset, publicly available at scrippsco2.ucsd.edu.

IPCC 2021 carbon budget: IPCC. Climate Change 2021: The Physical Science Basis. Cambridge University Press, 2021.

IEA renewable capacity: International Energy Agency. Renewables 2023. Paris: IEA, 2023.

Recommended Reading

The books listed here are those that have most directly influenced this work and that offer the most accessible and intellectually rewarding entry points into the disciplines it draws on.

Metamorphic Biology and Developmental Science

Carroll, Sean B. Endless Forms Most Beautiful: The New Science of Evo Devo. W.W. Norton, 2005. The definitive popular account of evolutionary developmental biology. Carroll's account of how a conserved genetic toolkit produces radically different body plans is essential background for Chapters One through Three.

Zimmer, Carl. Parasite Rex: Inside the Bizarre World of Nature's Most Dangerous Creatures. Free Press, 2000. Zimmer is among the most gifted science communicators working today; any of his books serve as models of responsible cross-disciplinary writing.

Cosmology and Physics

Krauss, Lawrence. A Universe from Nothing. Free Press, 2012. A clearly argued account of cosmic origins and stellar nucleosynthesis.

Rovelli, Carlo. Seven Brief Lessons on Physics. Allen Lane, 2014. The shortest serious introduction to modern physics available, with genuine philosophical depth.

Penrose, Roger. The Road to Reality. Jonathan Cape, 2004. Demanding but rewarding: the most rigorous account of physical reality available to non-specialist readers willing to do serious work.

Neuroscience and Psychology

Taylor, Jill Bolte. My Stroke of Insight. Viking, 2008. Essential reading for Chapter Seven.

Kahneman, Daniel. Thinking, Fast and Slow. Farrar, Straus and Giroux, 2011. Indispensable for understanding the psychological dimensions of metamorphic experience.

Bonanno, George. The Other Side of Sadness. Basic Books, 2009. The best evidence-based account of grief and resilience available for general readers.

Kessler, David. Finding Meaning: The Sixth Stage of Grief. Scribner, 2019. A practitioner's extension of the Kübler-Ross framework that adds genuine psychological depth.

Philosophy and Process Thought

Parfit, Derek. Reasons and Persons. Oxford University Press, 1984. The most rigorous philosophical treatment of personal identity. Part Three is directly relevant to Chapter Nine.

Whitehead, Alfred North. Process and Reality. Macmillan, 1929. The foundational text of process philosophy. Demanding but foundational.

Nagel, Thomas. Mind and Cosmos. Oxford University Press, 2012. A challenging argument that materialist accounts of consciousness are insufficient. Worth reading for the controversy it provoked as much as for its arguments.

History and Cultural Transformation

Ward-Perkins, Bryan. The Fall of Rome and the End of Civilization. Oxford University Press, 2005. The best account of the real costs of the Roman dissolution.

Kuhn, Thomas. The Structure of Scientific Revolutions. University of Chicago Press, 1962. The most influential account of intellectual paradigm shift, itself a metamorphic concept.

Diamond, Jared. Collapse: How Societies Choose to Fail or Succeed. Viking, 2005. A broad comparative survey of civilizational collapse and survival with direct implications for Chapter Sixteen.

Contemplative Traditions and Practice

Chödron, Pema. When Things Fall Apart: Heart Advice for Difficult Times. Shambhala, 1997. One of the most practically useful books for anyone in a chrysalis phase.

Turner, Victor. The Ritual Process: Structure and Anti-Structure. Aldine, 1969. The foundational anthropological account of liminality.

Rilke, Rainer Maria. Letters to a Young Poet. Translated by M.D. Herter Norton. W.W. Norton, 1934. The instruction to live the questions has informed the spirit of this book throughout.

The Future of Transformation

International Energy Agency. World Energy Outlook 2023. Available at iea.org. The most authoritative annual assessment of the global energy transition.

Tegmark, Max. Life 3.0: Being Human in the Age of Artificial Intelligence. Knopf, 2017. A thoughtful assessment of the implications of advanced AI with useful frameworks for Chapter Fifteen.

World Economic Forum. The Future of Jobs Report 2023. Available at weforum.org. Essential empirical grounding for the discussion of AI and labor markets.

About the Author

Calyx Drevorn is a researcher and writer whose work explores the intersections of biological science, cosmology, depth psychology, and philosophy. Trained in the history and philosophy of science, Drevorn has spent two decades investigating the patterns that recur across scales of natural organization, with particular attention to the dynamics of radical change and transformation.

The Silent Unfolding is the product of seven years of research across the disciplines of developmental biology, astrophysics, quantum mechanics, cognitive neuroscience, cultural anthropology, and process philosophy.

Drevorn lives and writes in southern England, in proximity to the fields and woodland edges where, on an autumn evening some years ago, a moth completed its emergence and flew into the dark.

www.ingramcontent.com/pod-product-compliance
Lightning Source LLC
LaVergne TN
LVHW090410210926
848480LV00002B/518

* 9 7 9 8 2 3 3 9 4 6 4 7 9 *